MW00395694

INSTANT
MATHEMATICS

Portable Press
An imprint of Printers Row Publishing Group
10350 Barnes Canyon Road, Suite 100, San Diego, CA 92121
www.portablepress.com • mail@portablepress.com

Printers Row Publishing Group is a division of Readerlink Distribution Services, LLC.
Portable Press is a registered trademark of Readerlink Distribution Services, LLC.

Correspondence regarding the content of this book should be addressed to Portable Press, Editorial Department, at the above address. Author and illustration inquiries should be addressed to Welbeck Publishing Group, 20–22 Mortimer St, London W1T 3JW.

Portable Press
Publisher: Peter Norton
Associate Publisher: Ana Parker
Editor: Stephanie Romero Gamboa
Senior Product Manager: Kathryn C. Dalby
Produced by Welbeck Non-fiction Limited

Library of Congress Control Number: 2019945898

ISBN: 978-1-64517-055-6

Printed in Dubai

23 22 21 20 19 1 2 3 4 5

All illustrations provided by Adam Cunningham and John Ringland, Noun Project & public domain sources.

INSTANT
MATHEMATICS

KEY THINKERS, THEORIES, DISCOVERIES, AND CONCEPTS EXPLAINED ON A SINGLE PAGE

PAUL PARSONS AND GAIL DIXON

San Diego, California

CONTENTS

ANCIENT

ROMAN TO MIDDLE AGES

EARLY MODERN

NINETEENTH CENTURY

TWENTIETH CENTURY AND BEYOND

INTRODUCTION

*It's difficult to imagine a world without mathematics.
No numbers, no rules of geometry, algebra, or logic—
no way of quantifying the world.*

The first evidence we have for the human understanding of mathematics dates back tens of thousands of years. The Lebombo bone, named after the Lebombo Mountains in Africa where it was discovered by archeologists in the 1970s, is a lower leg bone from a baboon. It's special because, deliberately cut into its length, are twenty-nine notches, believed to be tally marks. The bone's owner was keeping count of something—perhaps hunting kills, the members of an enemy war party, or simply the passing of days.

The bone has been radiocarbon dated to around the year 40,000 BC, making it the world's oldest mathematical artifact.

Counting is the most primitive example of mathematics. And it was clearly something humans *discovered* about the world, rather than invented. The number of apples in a basket obeys the rules of arithmetic. So, if I add one apple, the number in the basket increases by one. This has always been the case, whether people were aware of the fact or not.

The same is true of geometry. In around 300 BC, Greek philosopher Euclid wrote his book *Elements*, in which he set out the fundamental postulates describing the nature of circles, straight lines, and angles. These were rules of logic and inference, formulated to describe the world we live in. They constituted a discovery about that world, not an invention.

And yet, in the past, some very clever people have taken the opposite view: that mathematics is nothing more than a construct of the human mind. "The mathematics is not there till we put it there," declared the great twentieth-century English astronomer Sir Arthur Eddington.

That's certainly true in some cases. For example, "imaginary numbers" are a construct designed to make tangible the seeming impossibility of taking the square root of a negative number. Choose any ordinary number you like and square it, and the answer will *always* be positive. So how can you ever take a meaningful square root of something negative? Italian mathematician Rafael Bombelli came up with the answer in the sixteenth century. By denoting $\sqrt{-1}$ by the symbol i, he was able to express arbitrary square roots of negative numbers as a multiple of i, which could be manipulated using the same rules

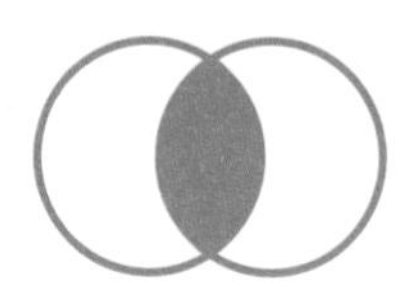

of algebra that governed ordinary numbers. It might sound like a ridiculous thing to do, but, today, imaginary numbers are a mainstay of wave theory, quantum mechanics, and data analysis.

Mathematical proofs are another example, where establishing the validity of a mathematical statement can entail a good deal of creativity on the part of the mathematician. For example, take Fermat's Last Theorem: the notion that the equation $x^n + y^n = z^n$, where x, y, z, and n are all whole numbers, has no solutions for $n > 2$. It took almost four hundred years to arrive at a watertight proof of this, which was ultimately found only thanks to a substantial feat of ingenuity on the part of the British mathematician Andrew Wiles.

From this perspective, the best mathematicians are architects with a creative vision that is at least the equal of their technical expertise. As Albert Einstein himself put it: "Logic will get you from A to B. Imagination will take you everywhere." And the insights gained as pioneering minds charted unknown mathematical territories were inevitably fed back to expand humanity's broader grasp of the subject.

In this book, we present a cheater's guide to the essentials of modern mathematics. Starting from the Lebombo bone and working forward in a (very roughly) chronological order, we lay out the major principles and discoveries, and detail the lives of those who made them.

Following on from the understanding of basic numbers and arithmetic, humans discovered the concepts of geometry, encompassing trigonometry, the calculation of areas and volumes, and Pythagoran theorem relating the three side lengths of a right-angled triangle. In recent centuries, geometry has grown to encompass curved spaces and *topology*—which governs how points in a space are connected to one another. A ring doughnut (a surface with a hole through it) has a very different topology to a sphere, for example.

There's evidence that *algebra*—the replacement of numerical quantities with symbols, and their subsequent manipulation to solve equations—was known

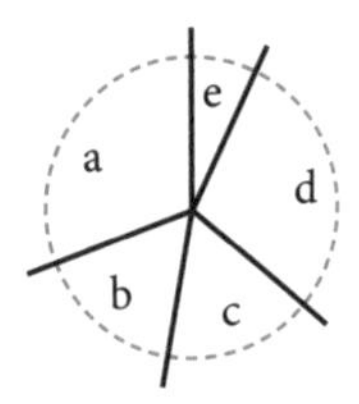

to Babylonian mathematicians as far back as 1800 BC. It was refined further during the European Dark Ages by mathematicians in Asia.

In the seventeenth century, polymaths Isaac Newton and Gottfried Leibniz developed calculus, used for inferring the rate of change of a numerical quantity, and that has since found applications in almost every branch of mathematics, as well as in theoretical physics.

At around the same time, a coterie of European mathematicians (who, it must be said, were also fond of supplementing their income by gambling) laid down the foundations of probability theory: the mathematical laws governing the likelihood of random events. From this grew statistics: the art of extracting information from data in the presence of random noise—converting messy, real-world observations into hard-and-fast knowledge, against which scientific theories can be tested and refined.

Mathematical reasoning also gave rise to computing. The number system we're most used to dealing with is base 10, built from the numerals 0–9. But in the second century BC, the Indian mathematician Pingala was the first to contemplate binary, or base 2 numbers, constructed from just 0 and 1. Binary later became the language of information. And in the nineteenth century, the desire to process this information in an automated way led to the invention of mechanical computers. These were followed by the room-sized behemoths used for codebreaking during the Second World War, and most recently by those tiny yet powerful information processors that go everywhere with us inside our mobile phones.

Today, mathematical principles form our toolkit for understanding everything—from quantum physics and the birth of the universe, to the stock market, to trivialities such as managing the weekly grocery budget. Indeed, as the owner of the Lebombo bone may well have realized, we live in a world that is fundamentally numerical.

COUNTING

Mathematics began with counting, the simplest use of numbers. Early people would use their fingers and thumbs, before moving on to making marks and the using tokens.

35,000 BC Bones have been found dating to this period that have tally marks scratched into them, possibly to measure elapsed time or numbers of animals

C.**4000 BC** The Sumerians of ancient Mesopotamia count using baked clay tokens marked with symbols. These are used for official accounting and were placed into clay "envelopes" to prevent tampering

C.**3000 BC** In Egypt, the cubit measure is introduced for building. It is defined as the length of a man's arm from elbow to fingertips

C.**3000 BC** The Egyptians begin using the earliest known base 10 system. Numbers are represented in hieroglyphs; for instance, 100 is a coil of rope

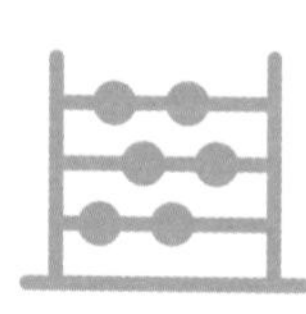

2700 BC The Sumerians use an early form of abacus, built according to a base 60 counting system

XIV

C.**900 BC** Evidence of the usage of Roman numerals (I, V, L, C, D and M) dates to this period. They were unwieldy numerically, so most counting was done on an abacus

C.**AD 500** Indian mathematicians devised a system where the numbers 1 to 10 were represented in symbols, the forerunner of today's decimal counting system

0

C.**AD 620** Zero was defined as a number by the Indian scholar Brahmagupta

123
456
789

C.**12th century** Use of the Hindu-Arabic numeral system (numerals 1 to 10) spread in Europe, popularized by the Italian mathematician Leonardo Fibonacci

15th to 16th centuries The modern numerals of our decimal system are by now widely used and incorporated into early typesetting

ARITHMETIC

The oldest and most important branch of mathematics, arithmetic was developed by the ancient Babylonians to facilitate trading, building, and record-keeping.

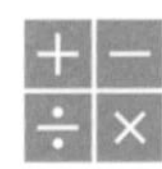

EARLY ACCOUNTANTS

From evidence, it's clear that the Sumerians of **Mesopotamia** had good understanding of the **four arithmetic functions**. From around 4000 BC, they counted using **clay tablets** and **stones on lengths of string**, like beads.

CALCULATING MINDS

Babylonian tablets dating back two thousand years have been found with early workings in **arithmetic**, **geometry**, and **algebra**. These include **multiplication tables**, **fractions**, and **quadratic and cubic equations**.

GIFTED GREEKS

The Greek word for number is ***"arithmós"*** and in 600 BC, the **Pythagorean school** determined arithmetic as one of the cornerstones of mathematics. The **ancient Greeks** elevated the age-old study of arithmetic problems to new heights, developing **geometry** and applying it to **astronomy**, **engineering**, and **design**.

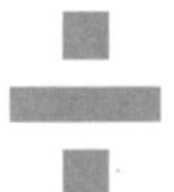

DIVIDED WE STAND

In **1597**, English mathematician **Henry Briggs** set out the modern method for **long division**, which is **still taught in schools today**.

MEDIEVAL MATHEMATICIANS

During the **Middle Ages**, arithmetic was taught in **European universities** as one of the **seven liberal arts**, alongside **grammar**, **logic**, **rhetoric**, **geometry**, **music**, and **astronomy**.

ADDING MACHINE

In 1948, **Curt Herzstark** invented a **handheld mechanical calculator**, which was replaced in the 1970s by **digital pocket technology** devised in **Japan**. Children in mathematics exams count their blessings.

THE RHIND PAPYRUS

This **Egyptian scroll**, made c.1550 BC, is **one of the earliest known mathematical texts**. Its eighty-four problems in **arithmetic**, **geometry**, and **volume** offer fascinating insight into ancient Egyptian number skills. The scroll is on display in the **British Museum**.

MAGIC SQUARES

This form of recreational mathematics dates back to ancient China and is a precursor of Sudoku and the Rubik's Cube. All are believed to be an excellent form of "brain gym."

8	1	6
3	5	7
4	9	2

This is an example of a **magic square**—a grid of positive whole numbers such that the sum of each row, column, and diagonal adds up to the **"magic constant,"** in this case 15.

HIGHER SQUARES

Euler found a fourth order square, each component of which is itself a squared number.

68^2	29^2	41^2	37^2
17^2	31^2	79^2	32^2
59^2	28^2	23^2	61^2
11^2	77^2	8^2	49^2

There are also "bi-magic" squares—magic squares that remain magic after their numbers are squared. And there are "tri-magic" squares (that stay magic after squaring or cubing their numbers), and even "magic cubes"—3-D blocks of numbers that must sum to the same magic constant along all rows, columns and "pillars," and on the four diagonals.

COMPETITION TIME

It's still unknown whether some types of exotic magic square are possible or not, and prizes are available for solutions. (See www.multimagie.com.)

ORDER!

A magic square with side length n is said to be of **"order n."** The above example is of order 3. In 1770, Swiss mathematician **Leonard Euler** became the first to construct a square of order 4.

The largest magic square was found in 2012, and is of order 3,559. It was 65 feet across when printed out.

MATH MYSTICISM

Scholars were beginning to toy with magic squares around 4,800 years ago in China. They were originally believed to have mystical properties. Magic squares indicate that early people experimented with arithmetic and understood that mathematics was far more than the sum of its parts! In the Lo Shu square (right) the numbers are given by the sum of the dots in each figure.

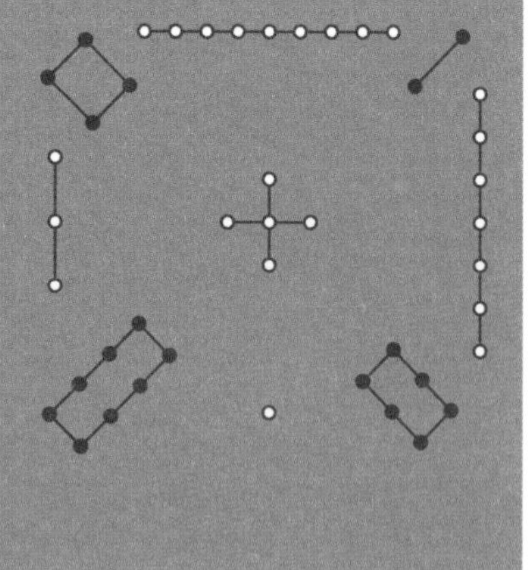

FRACTIONS

Taken from the Latin word fractio, *meaning to break, fractions are used constantly in everyday life. It's believed that the ancient Egyptians invented them to apportion taxes.*

FRACTIONS EXPLAINED

Fractions are a way to **represent portions of a whole number**. For example, $z = x/y$ means x parts of a possible y. If y is a whole-number multiple of x, then z is a whole number.

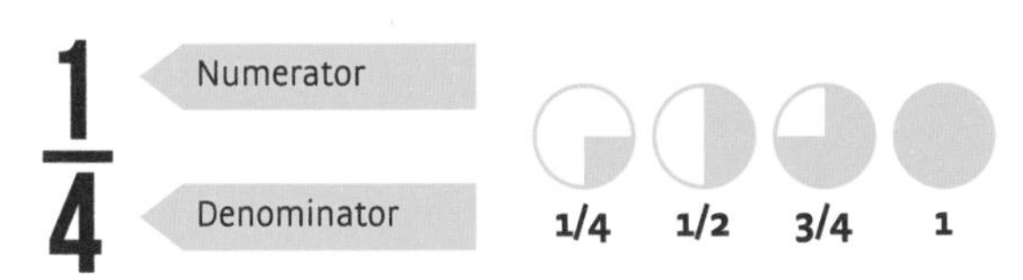

ADDING AND SUBTRACTING

Adding or subtracting fractions with the **same denominator** is easy. Just add up or subtract the numerators on the top row. For example,
$3/5 + 1/5 = 4/5$

But if the fractions do not have the same denominator, you need to calculate the **lowest common denominator**. For example, consider
$1/2 + 1/3 = ?$

The lowest common denominator of 2 and 3 is 6. Six is a multiple of 2 and 3, and is their lowest common denominator. So, converting both fractions to the lowest common denominator, gives
$1/2 \times 3/3 = 3/6$
$1/3 \times 2/2 = 2/6$

Now you can **add the numerators together**:
$3/6 + 2/6 = 5/6$

MULTIPLYING FRACTIONS

Simply **multiply the top and bottom numbers** in the fraction. For example,
$1/2 \times 4/5 = 4/10$.

Simplifying the fraction, in this case dividing top and bottom by 2, gives 2/5.

DIVIDING FRACTIONS

To divide fractions, **turn the second fraction upside down and multiply**. For example,
$1/2 \div 1/6 = 1/2 \times 6/1$.

Simplifying then gives the answer 3.

DECIMALS AND PERCENTAGES

Fractions, decimals, and **percentages** are all **equivalent** ...

Fraction	Decimal	Percent
1/2	.50	50%
1/3	.333	33.3%
2/3	.666	66.6%
1/4	.25	25%
3/4	.75	75%
1/5	.20	20%
2/5	.40	40%
3/5	.60	60%
4/5	.80	80%
1/6	.166	16.6%
1/8	.125	12.5%
1/10	.10	10%
1/12	.0833	8.3%

Gold is calculated in **carats**: 24-carat is pure gold, 18-carat is 18/24 (or 3/4); therefore it is 75 percent gold.

FRACTIONS IN EVERYDAY LIFE

Each minute is a fraction of the hour, so we use the terms "half an hour" and "a quarter to."

In **photography**, **shutter speed** is measured in **fractions of a second**.

TRIANGLES

The mathematical laws governing the geometry of triangles were cast on a solid footing by the ancient Greeks, most notably Euclid.

TYPES OF TRIANGLE

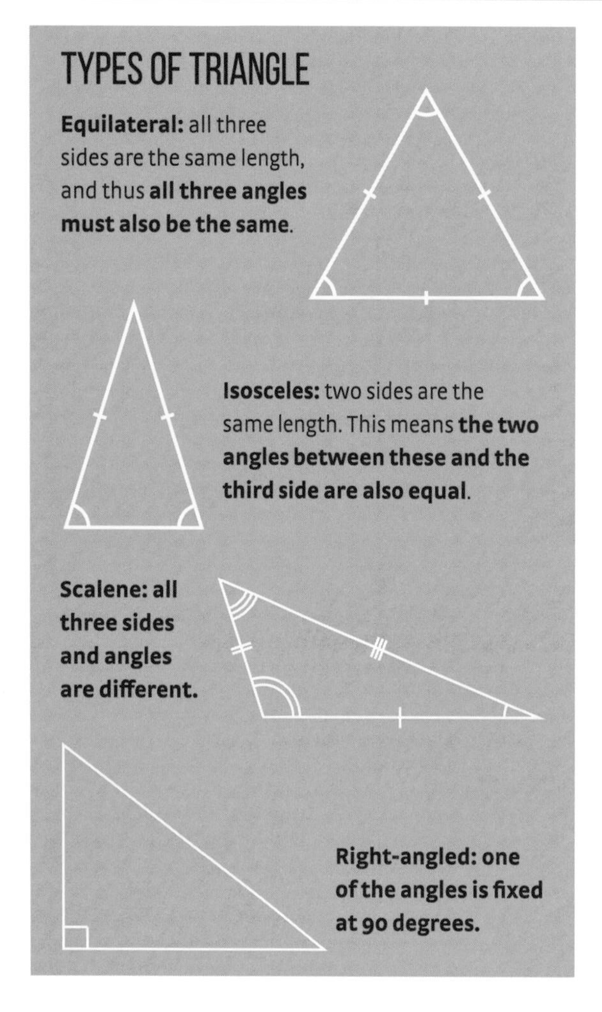

Equilateral: all three sides are the same length, and thus **all three angles must also be the same.**

Isosceles: two sides are the same length. This means **the two angles between these and the third side are also equal.**

Scalene: all three sides and angles are different.

Right-angled: one of the angles is fixed at 90 degrees.

TRIANGULATION

The **Greek philosopher Euclid** showed that the **internal angles of a triangle must sum to 180 degrees**.

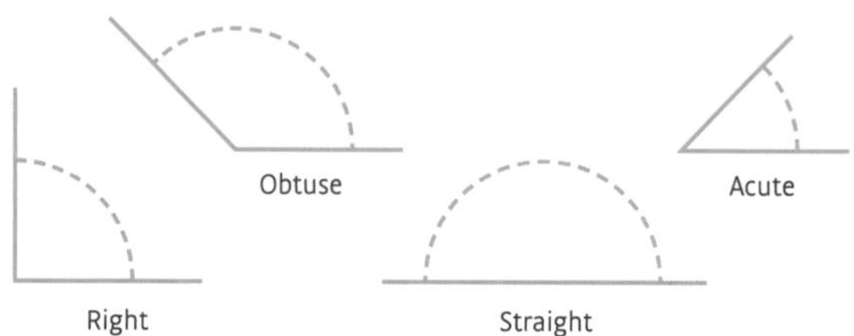

Angles greater than 90 degrees are called **"obtuse,"** while those smaller than 90 degrees are **"acute."**

Pythagoras showed that **the longest side squared is equal to the sum of the squares of the other two—for a right-angled triangle**. Generally, the longest side is shorter than the sum of the other two.

Two triangles are called **"similar"** if they are **the same shape but different sizes**. If they are the **same size and shape**—that includes **mirror images**—then they are **"congruent."**

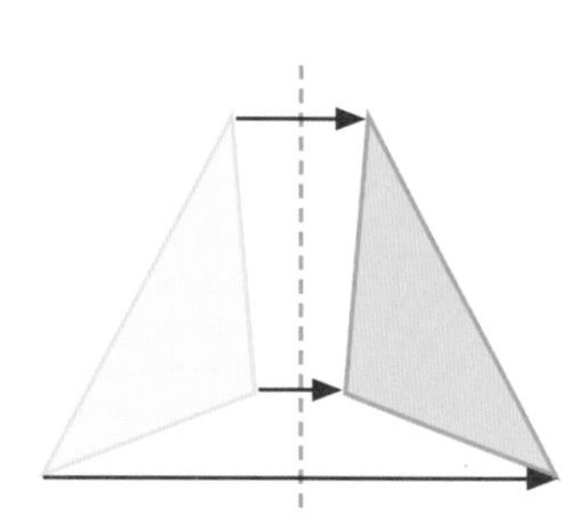

AREA

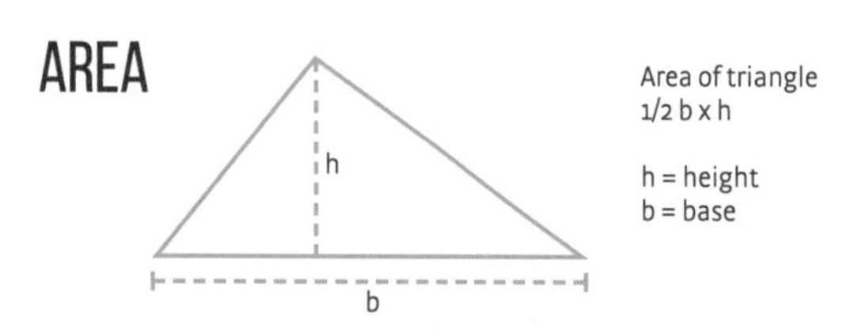

The **area of a triangle** can be calculated as **one half its base length multiplied by its vertical height**.

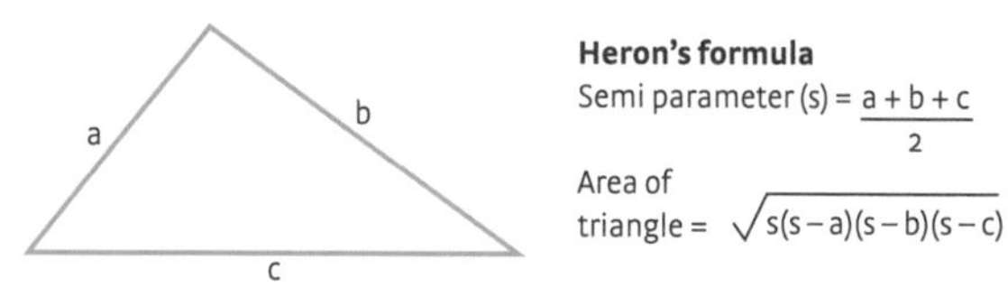

Greek mathematician Hero of Alexandria devised a more general method, known as **Heron's formula**. If the side lengths are a, b, and c, then the area is the square root of $s \times (s - a) \times (s - b) \times (s - c)$, where $s = \frac{1}{2} \times (a + b + c)$.

THALES OF MILETUS

Thales was the first scholar to use deductive reasoning in mathematics and the first known scientist to make a mathematical discovery.

THE WORLD'S FIRST MATHEMATICAL DISCOVERY

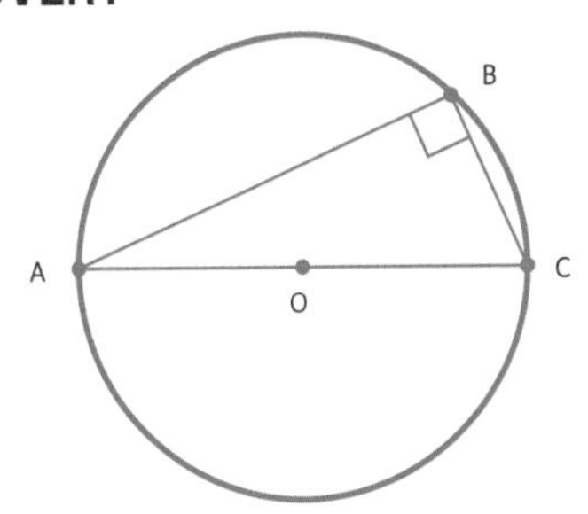

Thales was fascinated by **geometry**. This led to the world's **first recorded mathematical theorems**. The most famous—called **Thales's theorem**—describes **right-angled triangles within semi-circles**. Thales deduced that if A, B, and C are defined points on a circle, and A–C forms the diameter, then the angle at B will always form a right angle.

THEOREMS GALORE

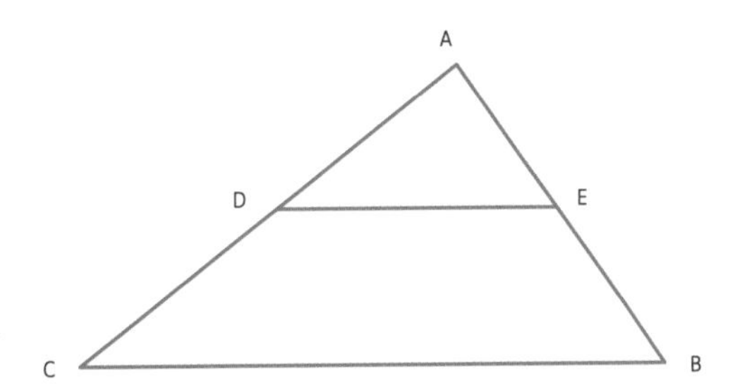

Thales also proved the **intercept theorem**—that if the lines CB and DE are **parallel**, then the ratios of CD/DA and BE/EA must be **equal**.

In addition, he showed that a **circle** is exactly **bisected by its diameter**, that the **two base angles of an isosceles triangle are equal**, and that **opposite angles formed by two crossing lines are equal**.

LIFE AND TIMES

c.624 BC: Thales was born in Miletus, Ionia, an ancient Greek province on the coast of modern-day Turkey. He was part of what Aristotle defined as the Ionian School, a group of influential philosophers and cosmologists.

Thales's family members were prosperous merchants, though he spurned the commercial life to ponder mathematics and astronomy.

Thales traveled to Egypt and ancient scholars reported that he measured the height of pyramids, by comparing his shadow to the shadow cast by the pyramid.

According to the Greek historian Herodotus, Thales predicted the solar eclipse of May 28, 585 BC.

It is thought that Thales died of heatstroke while watching the fifty-eighth Olympic Games, c.545 BC. He was around seventy-eight.

CRYPTOGRAPHY

The practice of formulating and cracking codes has become an increasingly mathematical endeavor. Today, it's essential for national defense and keeping our financial transactions secure.

ANCIENT

- **1500 BC** Mesopotamian clay tablets bear encrypted content
- **7th century BC** Greeks implement transposition ciphers
- **1st century BC** Roman emperor Julius Caesar uses a substitution cipher
- **800 AD** Frequency analysis developed by Arab mathematicians to break codes
- **1467** Leon Battista Alberti outlines polyalphabetic ciphers
- **1586** Mary Queen of Scots uses a substitution cipher in the plot to assassinate Queen Elizabeth I
- **1914–18** British Admiralty breaks a number of German naval codes
- **1939–45** The highly secure German Enigma and Lorenz codes—utilizing electromechanical cipher machines—are broken by cryptanalysts
- **1949** Claude Shannon publishes his revolutionary article *A Mathematical Theory of Cryptography*
- **1976** Public key cryptography is first outlined
- **1984** Absolutely secure "quantum cryptography" is proposed

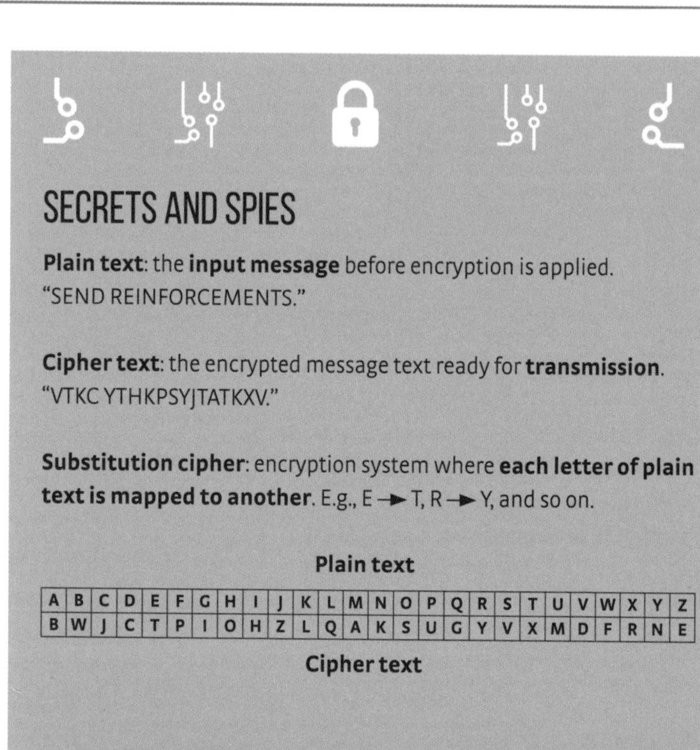

SECRETS AND SPIES

Plain text: the **input message** before encryption is applied. "SEND REINFORCEMENTS."

Cipher text: the encrypted message text ready for **transmission**. "VTKC YTHKPSYJTATKXV."

Substitution cipher: encryption system where **each letter of plain text is mapped to another**. E.g., E → T, R → Y, and so on.

Plain text

A	B	C	D	E	F	G	H	I	J	K	L	M	N	O	P	Q	R	S	T	U	V	W	X	Y	Z
B	W	J	C	T	P	I	O	H	Z	L	Q	A	K	S	U	G	Y	V	X	M	D	F	R	N	E

Cipher text

Transposition cipher: Encryption where plain-text characters are written into a grid and the grid cells are shuffled according to a fixed set of rules.

	1	2	3
1	B	E	S
2	T	C	O
3	D	E	S

Grid 1

→

	2	1	3
3	E	D	S
1	E	B	S
2	C	T	O

Grid 2

Frequency analysis: a way to crack substitution ciphers by noting how often each letter crops up in the cipher text. E.g., "E" is the most common letter in English, occurring 12.7 percent of the time. So if, say, 12.7 percent of the letters in the cipher text are "T"s then it's a fair bet that these decrypt to "E"s.

Polyalphabetic ciphers: frequency analysis can be thwarted by substituting from multiple alphabets used in rotation. This is essentially how the German Enigma code worked.

EXPONENTIATION

Exponentiation sits with the other arithmetic operations of addition, subtraction, multiplication, and division, and is shorthand for multiplying a number by itself many times.

WHAT IS IT?

Exponentiation is a mathematical operation that involves **raising one number**, *a*, to the **"power" of another**, *b*. In modern notation, this is written a^b. When *b* is an integer, it's equivalent to multiplying together *b* **factors** of *a*. So, for example, $a^4 = a \times a \times a \times a$.

2nd to 3rd century BC Euclid coins the term "power" to describe the square of a number

3rd century BC Archimedes discovers that $10^a \times 10^b = 10^{a+b}$

1544 The word "exponent" is coined by German mathematician Michael Stifel

1637 René Descartes, in his book *La Geometrie*, is the first to use modern exponential notation

1748 Leonhard Euler considers expressions in which the exponent itself is the variable, e.g., 2^x

INVERSES AND ROOTS

Negative powers correspond to the **"inverse" of a number**. So, $x^{-n} = 1/x^n$.

A **fractional power** denotes a **root**. So, if $y = x^2$, that means $x = y^{1/2} = \sqrt{y}$. Here, √ is the **radical symbol**, used to indicate the **square root** of *y*; that is, a number which when multiplied by itself equals *y*. It's also used for higher order roots, so $x^{1/n} = \sqrt[n]{x}$, or the nth root of x.

And anything raised to the power 0 is just 1.

Powers of 4

Power	Result	Name
-1	1/4	Inverse
0	1	NA
1/2	2	Square root
1	4	NA
2	16	Square
3	64	Cube
4	256	Fourth power

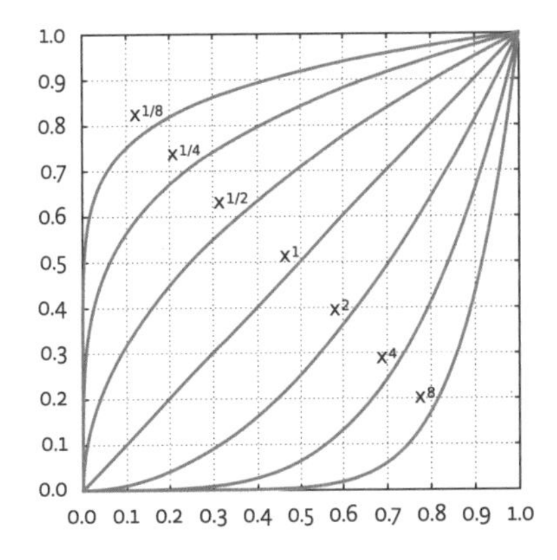

Powers and roots of numbers between 0 and 1

AREA AND VOLUME

From cookery to surveying to the physics of the cosmos, area and volume are essential mathematical tools for quantifying the world.

ANCIENT

2000 BC First calculation of the area of a triangle

260 BC Archimedes uses ϖ to calculate the area and volume of circles and spheres

AD 1020 Abul Wafa speculates on the volume of a paraboloid

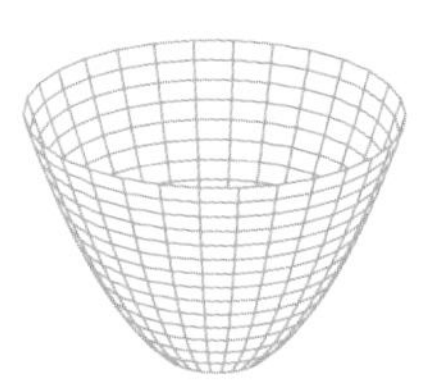

17th century Integral calculus is developed, offering a way to calculate the area and volume of arbitrary mathematical shapes

$\int$

AREA

A **1-D line** has a single, simple property: its **length**.

For a **2-D shape**, you can measure the length around its outer perimeter. But the shape also has a second property: its **area**. This quantifies how much two-dimensional surface it contains.

A simple **rectangle**, for example, has an area equal to **length × width**.

In the real world, area might be measured in **square feet**, or **acres**.

AREAS OF SHAPES

Square side length squared

Rectangle length × width

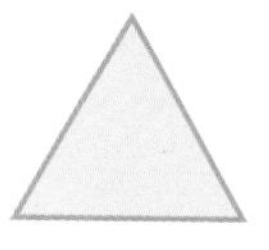

Triangle ½ × base length × height

Parallelogram base length × height

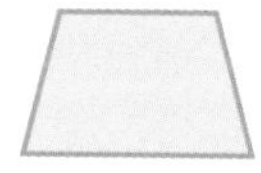

Trapezium ½ × base length × top length

VOLUME

Similarly, **3-D** objects have an outer surface area and an extra property: their **volume**. This tells you how much three-dimensional space they take up.

For example, a **cuboid** (a solid with rectangular faces that meet at right-angles to each other) has a volume equal to its **length × width × height**.

Volume is measured in units such as **cubic feet** or **pints**.

VOLUMES OF SOLIDS

Cube side length cubed

Cuboid length × width × height

Prism base area × height

Pyramid 1/3 × base area × height

Tetrahedron √2/12 × side length cubed

PYTHAGORAS

One of the first philosophers, the Ionian scholar Pythagoras made important contributions to mathematics, astronomy, and the theory of music.

Born: Samos, Greece, 570 BC.

Traveled in: Egypt, Persia.

Lived in: Croton, southern Italy.

Passions: philosophy, music, mathematics, politics, religion, metaphysics.

Died: approx. 495 BC.

PYTHAGORAN THEOREM

This famous **mathematical relationship** has been a cornerstone of **construction** and **measurement** for millennia.

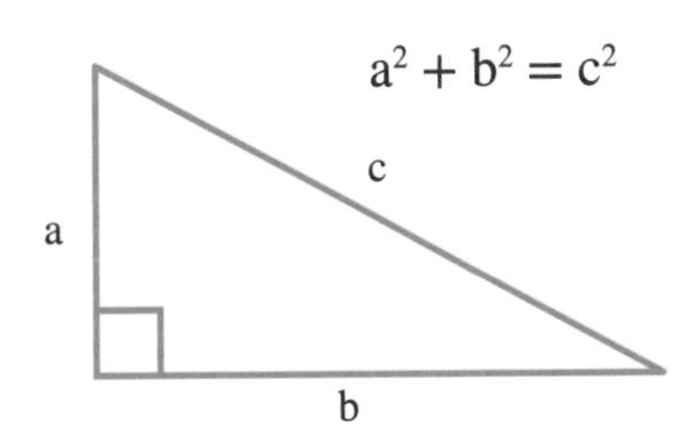

Pythagoras realized that **the square of the hypotenuse** (the side of a triangle opposite the right angle) is **equal to the sum of the squares of the other two sides**.

MYSTICISM

Pythagoras believed in **metempsychosis**—the **immortality of the soul** and its **reincarnation** after death.

HARMONY OF THE SPHERES

Astronomy fascinated Pythagoras, and he believed that **stars** and **planets** moved according to mathematics corresponding to **musical notes**, thus producing an **audible planetary symphony**.

PYTHAGOREANISM

Pythagoras inspired a **cult-like following** among his students and they lived as an **intellectual community**. Strict codes governed **behavior**, **dress**, and **diet**. Unusually for the time, Pythagoras welcomed **female students**.

SACRED FIGURES

Pythagoras believed that all things are **made of numbers**, and he attributed **sacred significance** to them.

THE PERFECT 10

Pythagoras is believed to have devised the **tetractys**, which is a **triangle** consisting of ten points arranged in four rows, containing one, two, three, and four points respectively.

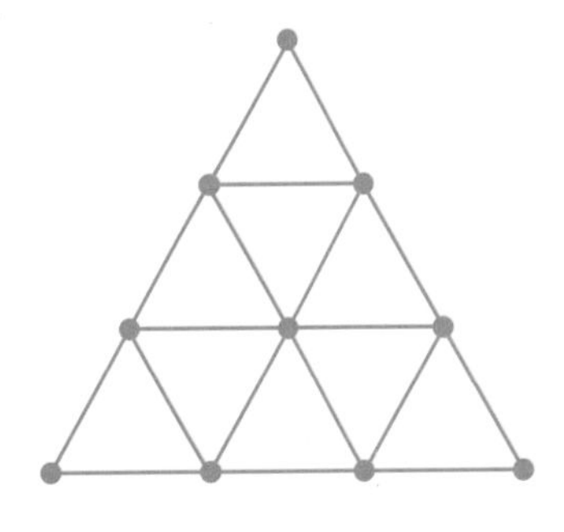

PYTHAGORAN THEOREM

One of the defining rules of Euclidean geometry is the Pythagoran theorem, which connects the lengths of the sides of a right-angled triangle.

TAMING TRIANGLES

The theorem states that given a right-angled triangle, the **sum of the squares** of the **two shorter sides** is **equal to the square of the longest side**—the **hypotenuse**, as it's known.

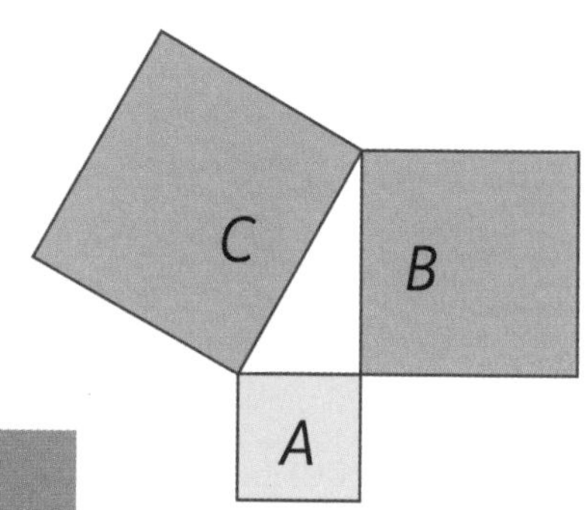

DID YOU KNOW?

The Pythagoran theorem for a right-angled triangle is usually stated in terms of the areas of squares of sizes equal to the triangle's side-lengths, but it actually works with **any geometric shape**. For example, stick **regular pentagons** to each side of a right-angled triangle and the area of the pentagon on the longest side still equals the sum of the areas of those on the two shorter sides.

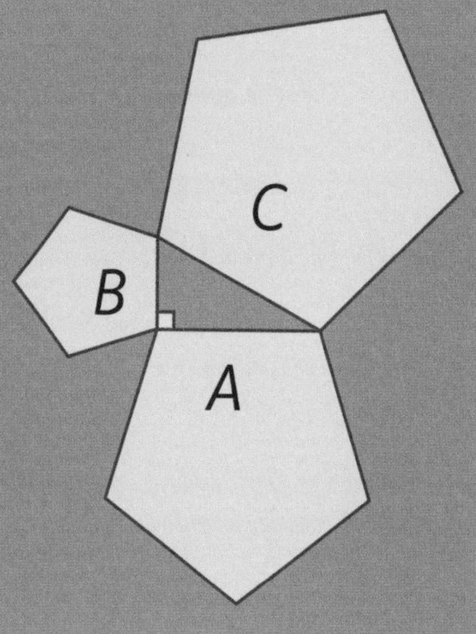

WHODUNNIT?

Although named after the Greek philosopher Pythagoras, who lived in the sixth century BC, the theorem is believed to have been known to the ancient **Mesopotamians**—1,200 years earlier. Pythagoras has been credited with the first proof—deriving the theorem from basic principles—although even the evidence for this has long been lost. A proof does feature, however, in **Euclid's** seminal work ***Elements***.

PYTHAGOREAN TRIPLES

There are particular **whole-number lengths** for the two shortest sides of the triangle for which the theorem also gives a whole-number length for the hypotenuse. And these special cases are known as the **Pythagorean triples**.

There are five sets of triples with values all less than 50. They are:

Side 1	Side 2	Hypoteneuse
3	4	5
5	12	13
7	24	25
8	15	17
9	40	41

IRRATIONAL NUMBERS

The irrationals are the ugly ducklings of the numerical world. Their discovery shattered the neat order of ancient Greek mathematics.

NUMERICAL ODDBALLS

Any number that can be written as a simple **fraction**, or **"ratio"**—that is, a/b where a and b are both **integers** (whole numbers)—is known as "rational." Conversely, any number that can't be written in this way falls into the mathematical category of **"irrational."**

EXAMPLES

Irrational numbers are never-ending and could in principle be written to an **infinite number of decimal places**. Not only that, there's **no predictable, repeating pattern to their digits**.

Examples of rational numbers include the **geometrical constant** π (= 3.14159...), the **Euler's number** e (= 2.71828...), and **square roots of any number not a perfect square**, such as √2 (= 1.41421...).

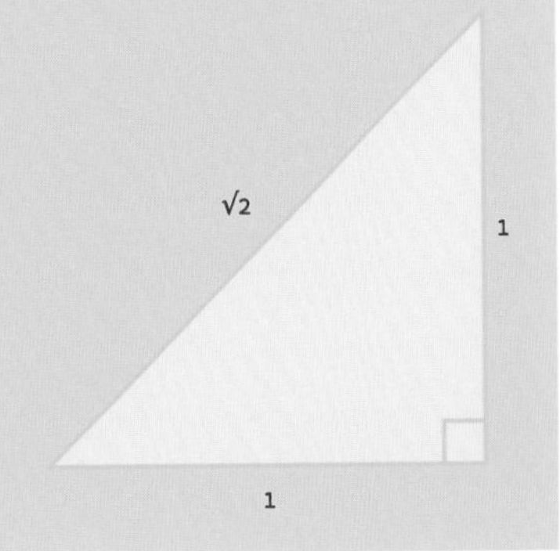

HIPPASUS

Irrational numbers were discovered by **Hippasus**, a follower of **Pythagoras**, in the sixth century BC. That mathematics should behave in such an **imperfect and unpredictable way** was shocking to the ancient Greeks. **Some authors even have Pythagoras drowning his protégé** for this outrageous finding.

6th century BC Hippasus discovers irrational numbers

4th century BC Eudoxus of Cnidus proves irrationality of square roots up to 17

5th century AD Aryabhata calculates π to four decimal places

9th to 10th century Abu Kamil Shuja' ibn Aslam considers irrational numbers as the solutions to equations

1768 Johann Heinrich Lambert suggests e and π are transcendental as well as irrational

HYPATIA

Famed for being the first known female mathematician, Hypatia was a pioneer who was brutally murdered for her beliefs.

WHO WAS HYPATIA?

Hypatia was born around AD 350, in **Alexandria, Egypt**.

She was **daughter of Theon of Alexandria**, an eminent Greco-Roman **mathematician** and **astronomer**.

Her name in Greek means **"supreme."**

According to the writer **Damascius**, Hypatia was **"exceedingly beautiful"** and chose to remain a virgin all her life.

EARLY LEARNING

Theon was a leading teacher and an adherent of **Neoplatonism**, a **religious and philosophical system** that developed in the **third century AD**. Hypatia's early years were spent within an inspiring intellectual environment.

PRESERVING KEY WORKS

Hypatia was dedicated to preserving Greek mathematical and astronomical works. She edited works by Archimedes. And she wrote commentaries on **Apollonius of Perga's** seminal work ***Conics*** and **Diophantus of Alexandria's *Arithmetica***.

A CAREER WOMAN

Alexandria was a prestigious center of learning. Hypatia taught **mathematics, astronomy,** and **philosophy** to **scholars from across the Mediterranean**. Her reputation grew in a career that spanned decades.
In a male-dominated world, she became one of its finest mathematicians and astronomers.

BRUTAL MURDER

Hypatia's **philosophical beliefs** and reputed political influence made her a target of hatred to some factions. This culminated in her **murder** in AD 415 by a band of **Christian zealots**, who dragged her into a church, **attacked her** with oyster shells, and **tore her limb from limb**.

A POWERFUL LEGACY

Hypatia is regarded as a **martyr of philosophy** and became a **powerful feminist symbol**. She inspired **writers, painters, philosophers,** and **scientists** for millennia.

POLYHEDRA

Polyhedra are 3-D solids made of polygons, which interlock to form edges and vertices. The laws governing their structure reflect the geometry of the space we live in.

BUILDING BLOCKS

Polygons are the building blocks of polyhedra. They are flat 2-D shapes composed of **straight-line segments** forming a **closed loop**. They can be **"regular"** (all sides the same length and inclined at the same angle), or **"irregular."**

ANATOMY OF A POLYHEDRON

The Swiss mathematician **Leonhard Euler** found that for **3-D polyhedra** with F faces, E edges, and V vertices, the formula

$$V + F - E = 2$$

must apply.

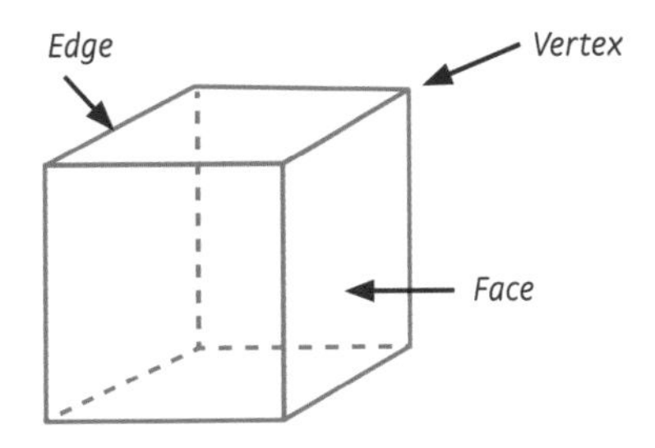

Euler's polyhedron formula holds for any 3-D shape that's **"simple"** (i.e., has no holes in it) and **"convex"** (i.e., is curved outwards all round like a ball).

One consequence is that **there can be no polyhedron with seven edges**.

PLATONIC SOLIDS

The **Platonic solids** are **"regular" polyhedra**—that is, all their faces are regular polygons with the same number of sides, and all the vertices have the same number of edges connecting to them. Applying these constraints to **Euler's formula** reveals that there **can only exist five regular polygons**, with four, six, eight, twelve, and twenty faces.

Board-game fans might recognize these as the bases for **multi-sided dice**—apart from the ten-sided die, which is a **pentagonal trapezohedron** (not a regular polyhedron, because its faces are not regular polygons).

Platonic solid	Faces	Face polygon
Tetrahedron	4	Equilateral triangle
Cube	6	Square
Octahedron	8	Equilateral triangle
Dodecahedron	12	Regular pentagon
Icosahedron	20	Equilateral triangle

IRREGULAR POLYHEDRA

A multitude of **other polyhedra** exist. For example, the **Archimedean solids** are each made from different regular polygons.

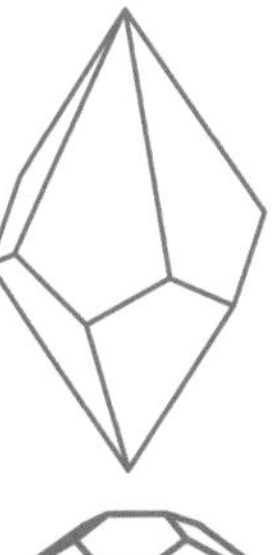

Perhaps the most famous of these is the **truncated icosahedron**—used for **soccer balls**.

PERFECT NUMBERS

Greek philosopher Euclid first described perfect numbers 2,300 years ago. Some scholars even believed their rare properties gave them a mystical significance.

WHAT IS A PERFECT NUMBER?

A perfect number is a **positive integer that is the sum of all its divisors, excluding itself** (see table). The number 6 is the first perfect number, since all the factors of 6—1, 2, and 3—also sum to 6.

Perfect number	Positive factors	Sum of all factors excluding itself
6	1, 2, 3, 6	6
28	1, 2, 4, 7, 14, 28	28
496	1, 2, 4, 8, 16, 31, 62, 124, 248, 496	496
8,128	1, 2, 4, 8, 16, 32, 64, 127, 254, 508, 1,016, 2,032, 4,064, 8,128	8,128

There are only **four perfect numbers under a million**. These are **6**, **28**, **496**, and **8,128**.

MYSTICAL MATHEMATICS

Perfect numbers have been **studied since ancient times** by scholars, who found them beguiling. **Pythagoras** believed that they had mystical properties.

In the fourth century BC, **Euclid** postulated that $2^{p-1}(2^p - 1)$ is a perfect number, and also an even integer, when $2^p - 1$ is a prime number. Here are some examples:

If p is 2: $2^1(2^2 - 1) = 2 \times 3 = 6$

If p is 3: $2^2(2^3 - 1) = 4 \times 7 = 28$

If p is 5: $2^4(2^5 - 1) = 16 \times 31 = 496$

If p is 7: $2^6(2^7 - 1) = 64 \times 127 = 8{,}128$

Understanding of perfect numbers led to **"Mersenne primes"** (prime numbers of the form $2p - 1$, where p is an integer), named after the **French monk Marin Mersenne** who discovered them.

In the eighteenth century, **Leonhard Euler** proved that any even perfect numbers can be found by using **Euclid's formula** (above). It's **unknown whether there are any odd perfect numbers**.

There are currently a total of **fifty-one perfect numbers known**.

PRIME NUMBERS

The existence of prime numbers has been known for thousands of years, but their properties are still not fully understood. They are one of the great curiosities of mathematics.

WHAT IS A PRIME?

Any whole number larger than 1, that can only be divided exactly by itself or 1, is a prime. So, 5 is prime because it can't be divided without remainder by any number other than 1 or 5. But 6 is not because, as well as 1 and 6, it can also be divided by 2 and 3.

The first twenty prime numbers are ...

2 3 5 7 11 13 17 19 23 29 31 37 41 43 47 53 59 61 67 71

There are **no even primes bigger than 2**. And the Greek philosopher **Euclid** proved that the **total number of primes is infinite**.

Today, **new prime numbers** are found using **computers**.

The **Riemann hypothesis**—an **unproven theorem** in **pure mathematics**—suggests that the **distribution** of prime numbers can be **predicted**.

APPLICATIONS

Prime numbers are essential to **other branches of mathematics**, as well as **cryptography**—and even **biology**...

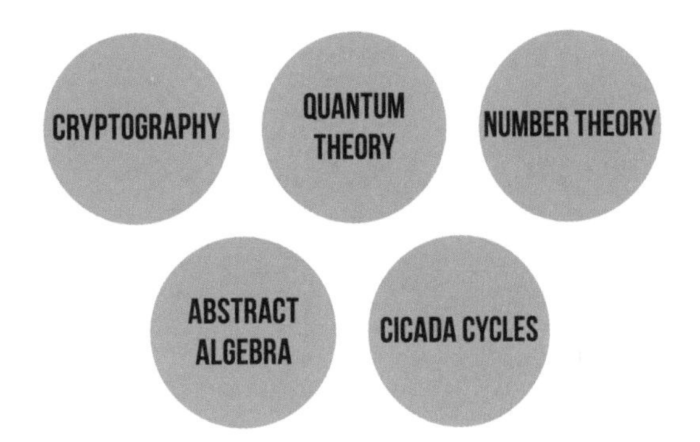

DID YOU KNOW?

Prime numbers were mentioned in the **Rhind Papyrus**, dating from 1550 BC.

The **largest known prime number** was discovered in December 2018, and takes the value $2^{82,589,933} - 1$.

There are many types of **"exotic prime"**—such as **Mersenne primes**, which are **always 1 less than a power of 2**, and **"sexy primes,"** which **come in pairs separated by 6**

MATHEMATICAL PROOF

Proof is the lifeblood of modern mathematics. Any theorem must be demonstrated as correct by a series of logical steps starting from established mathematical truths.

ANCIENT

- **5th to 6th century BC** Thales and Hippocrates prove early theorems in geometry
- **300 BC** Euclid pioneers proof by deduction from axioms
- **10th century AD** Development of algebraic proofs by Islamic scholars
- **1976** Kenneth Appel and Wolfgang Happel prove the four-color theorem
- **1994** Andrew Wiles proves Fermat's Last Theorem
- **2006** Grigori Perelman proves the Poincaré conjecture

WHAT IS IT?

Whether it's the **Pythagoran theorem** or the **Poincaré conjecture**, the **rules of math** are only as good as their **proofs**. Proofs take numerous forms, from simply demonstrating something **exists** (say, a solution to an equation), to proving **uniqueness** (for example, that a solution is the only solution), or proving a **general fact** (e.g., that $\sqrt{2}$ is irrational).

HOW?

Methods include **deduction** (where one step follows from another), **reductio ad absurdum** (proving that if a theorem isn't true then a **contradiction** results), and **induction** (where a result is **proven in one special case**, say $n = 0$, and then proven to hold in general when $n \rightarrow n + 1$).

The ability to **prove something 100 percent** true in mathematics stands in **stark contrast to the sciences**, where theories **can only ever be proven false**.

UNPROVEN PROBLEMS

There are some ideas in math that have **yet to be proven formally**. Most prominent of these are the six **Millennium Prize Problems**, set by the **Clay Mathematics Institute**, in New Hampshire. There's a **$1m prize** if you can solve any of these...

P vs NP: if a solution to a problem can be checked quickly, can it also be found quickly?

Hodge conjecture: that certain "homology classes" in an arbitrary space are algebraic.

Riemann hypothesis: that the non-trivial zeroes of the Riemann zeta function have a "real" component of ½.

Yang–Mills existence and mass gap: that the equations governing the strong nuclear force inside atomic nuclei can be solved.

Navier–Stokes existence and smoothness: prove the existence, or not, of smooth solutions to the Navier–Stokes equations of fluid dynamics.

Birch and Swinnerton-Dyer conjecture: tell whether elliptic curves have a finite or infinite number of rational solutions.

EUCLID

Euclid's book Elements *was long regarded as the definitive text on mathematics. Einstein described it as his "holy little geometry book."*

"EUCLID ALONE HAS LOOKED ON BEAUTY BARE."—EDNA ST VINCENT MILLAY (U.S. POET)

WHO WAS EUCLID?

Little is known of this **Greek scholar**, who was born in the fourth century BC and spent most of his life teaching and studying in **Alexandria, Egypt**.

His name in Greek translates to **"renowned"** or **"glorious."**

However, in antiquity, Euclid was rarely mentioned by name, being referred to rather as **"the author of *Elements*."**

Through time, he has become widely regarded as the **"father of geometry."**

He **died** in **mid-third century BC**.

ELEMENTS

Euclid's book ***Elements*** drew together all previous mathematical theory and teachings, including the work of **Eudoxus**, **Pythagoras**, and **Hippocrates**.

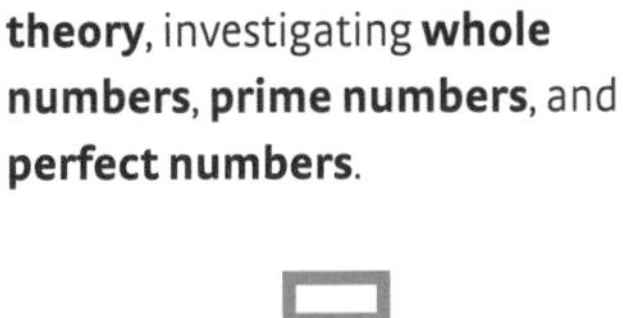

The book also featured one of the earliest treatments of **number theory**, investigating **whole numbers**, **prime numbers**, and **perfect numbers**.

It laid the ground work for **geometry**, the mathematics of shapes. Using just **drafting compasses** and a **straight line**, Euclid presented **hundreds of theorems**.

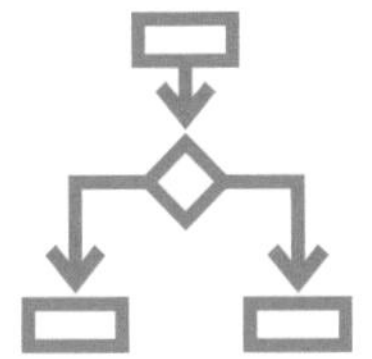

Euclid's method for finding the **largest common divisor of two numbers** is an early example of an "**algorithm**."

LEGACY

Modern, **advanced geometric ideas** such as **curved spaces** and **differential geometry** all derive from the early work of Euclid.

The European Space Agency's *Euclid* spacecraft, aimed at helping us to understand the **large-scale geometry of the universe**, launches in 2022.

EUCLIDEAN GEOMETRY

In 300 BC, Euclid collected all the existing knowledge of geometry—along with his own theorems and discoveries—and set them down as a thirteen-volume book, called Elements.

EUCLID'S AXIOMS

Euclid's geometric theorems were all derived from a small number of basic and largely **intuitive assumptions**, his **"axioms."**

1. You can draw a straight-line segment between **any two points**.
2. A straight line can be **extended infinitely** in either direction.
3. It's possible to **draw a circle** of any radius at any point.
4. All **right angles are equal**.

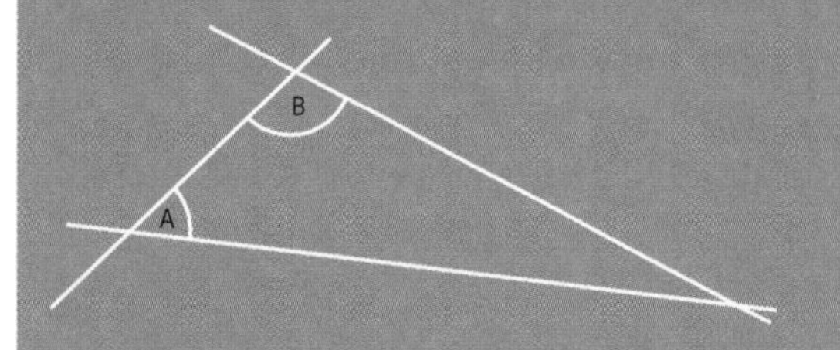

5. If two straight lines are crossed by a third line, then, if the internal angles formed add up to less than two right angles (180°), the lines will eventually cross. In other words, **parallel lines** (where the internal angles exactly equal (180°) **never meet**. This was known as Euclid's **"parallel postulate."**

LOOKING AT EUCLID

Euclid proved the theorems in *Elements* by **geometric construction**—that is, by drawing lines with a **straight edge** and **compasses**—and by **deduction** from his **axioms**.

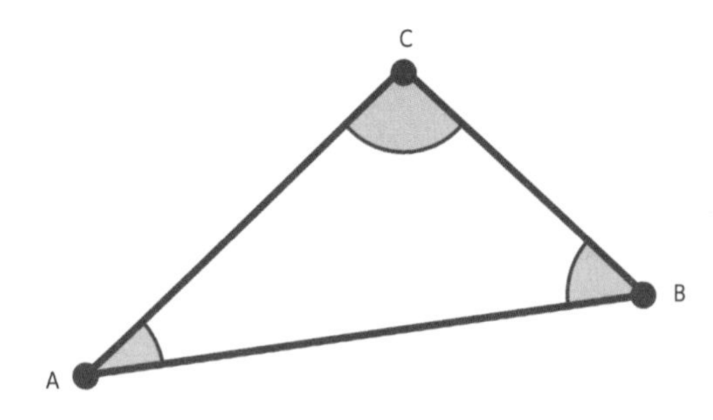

Angle A + Angle B + Angle C = 180°

Examples include the **interior angles of a triangle** (namely, that they all sum to 180°) and **Thales's theorem** (that a triangle drawn inside a circle, so that one side of the triangle is a diameter of the circle, must always be right-angled).

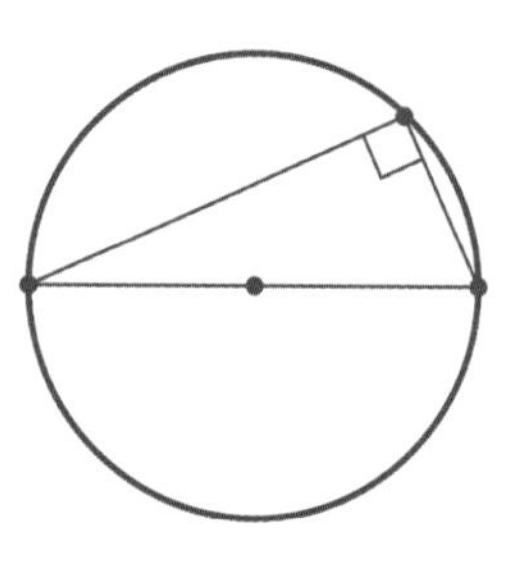

1650 BC Writing on geometry features in the Egyptian Rhind Papyrus

7th century BC Thales uses geometry to solve practical problems

300 BC Euclid's book *Elements* is published

1637 René Descartes's book *La Geometrie* marks the beginning of analytic geometry

1830 First results are published on non-Euclidean geometry

ALGORITHMS

Anyone who's written even the most basic computer program knows what an algorithm is—a specific set of instructions to automatically perform a task or solve a problem.

WHAT'S IN A NAME?

The word **"algorithm"** can be traced to the Persian scholar **Muhammad ibn Musa al-Khwarizmi**, whose name was Latinized by a translator in AD 825 to **Algoritmi**. This was later altered to **"algorithmus"** which, by the seventeenth century, became **"algorithm."** The word didn't take on its modern meaning until the nineteenth century.

ALGORITHMS IN DAILY LIFE

Recommendations: song or movie recommendations presented to you on platforms such as **iTunes** and **Netflix** use an algorithm to find **new selections** that are **comparable to your past preferences**.

Optimization: algorithms can find the **best solution from a multitude of possibilities**. It's how GPS finds the shortest route to a desired destination.

Searching: Google search and other algorithms can **scour the internet or a database** looking for the **best match to an input query**.

Sorting: when **Amazon** ranks your hits by **price** or **rating**, it's using an algorithm to **compare pairs of entries** in turn and then place the whole list in order.

EUCLID'S ALGORITHM

Probably the oldest example we have is Euclid's algorithm, which was published in Book VII of his ***Elements*** in 300 BC. The algorithm is **designed to find the greatest common divisor of two numbers**—that is, the largest value that both numbers can be divided by without remainder—which it **achieves by repeated subtraction of one number from the other**.

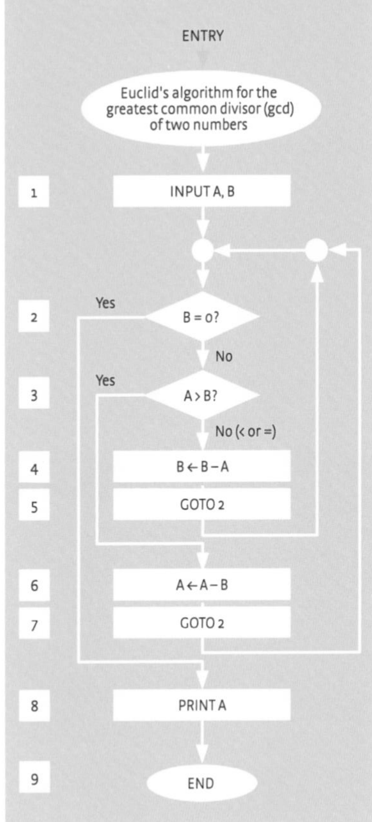

ANCIENT

PI

Pi, or π as it's usually written, is a fundamental constant of geometry, fixing the ratio of the circumference of a circle to its diameter.

WHAT IS π?

The modern value is:

3.14159265358979323846264338327950288419716939937510582097494459230781640628620899862803482534211706798214808651328230664709384460955058223172535940812848
8111745028410270193852110555964462294895493038196442881097566593344612847564823378678316527120190914564856692346034861045432664821339360726024914127372458
8700660631558817488152092096282925409171536436789259036001133053054882046652138414695194151160943305727036575959195309218611738193261179310511854807446237
9962749567351885752724891227938183011949129833673362440656643086021394946395224737190702179860943702770539217176293176752384674818467669405132000568127145
2635608277857713427577896091736371787214684409012249534301465495853710507922796892589235420199561121290219608640344181598136297747713099605187072113499999
9837297804995105973173281609631859502445945534690830264252230825334468503526193118817101000313783875288658753320838142061717766914730359825349042875546873
1159562863882353787593751957781857780532171226806613001927876611195909216420198938095257201065485863278865936153381827968230301952035301852968995773622599
4138912497217752834791315155748572424541506959508295331168617278558890750983817546374649393192550604009277016711390098488240128583616035637076601047101819
4295559619894676783744944825537...

That is a rough approximation, π is both **irrational** and **transcendental**, having an **infinite number of digits**. A pretty good rational approximation is 355/113.

CIRCLES

From the definition of ϖ, the **circumference of a circle** is just $C = 2\pi r$, where r is its radius. And its area is given by $A = \pi r^2$.

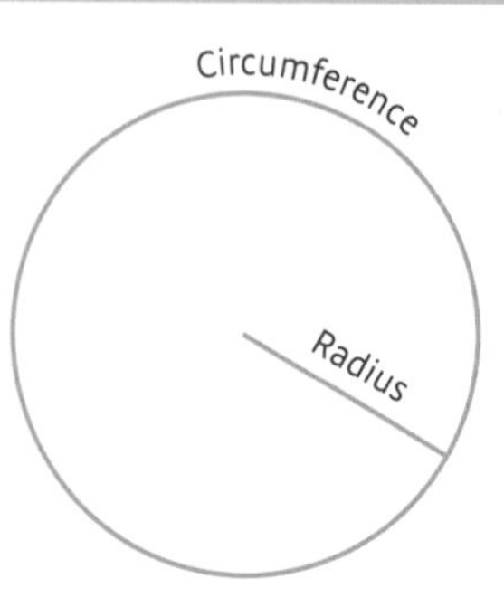

SOLIDS

Similarly, π features in the formulas for the **surface area and volume of three-dimensional solids that have circular cross sections.**

Solid	Surface area	Volume
Sphere	$4\pi r^2$	$\frac{4}{3}\pi r^3$
Cylinder	$2\pi r^2 + 2\pi rh$	$\pi r^2 h$
Cone	$\pi r(r + \sqrt{(h^2 + r^2)})$	$\frac{1}{3}\pi r^2 h$

- **1650 BC** First known estimate of π (~ 3.16) appears in the Egyptian Rhind Papyrus
- **3rd century BC** Archimedes approximates π by the fraction 22/7
- **1220** Fibonacci obtains a value of 3.1418 by approximating a circle as a many-sided polygon
- **1706** Welsh mathematician William Jones is the first to use the symbol π
- **1761** Swiss mathematician Johann Heinrich Lambert demonstrates that π is irrational
- **1882** German mathematician Ferdinand von Lindemann proves that π is also transcendental
- **2015** Rajveer Meena recites 70,000 digits of π from memory, a world record
- **2019** Google employee Emma Haruka Iwao uses a computer to calculate π to 31,415,926,535,897 (almost 31.5 trillion) digits, also a world record

ARCHIMEDES

Widely considered to be the "father of mathematics," Archimedes was one of the leading scientists of the ancient world.

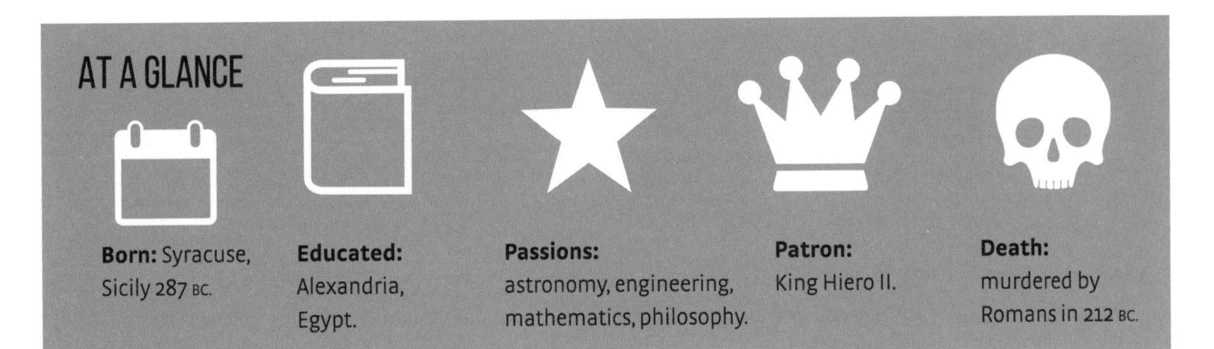

Born: Syracuse, Sicily 287 BC.

Educated: Alexandria, Egypt.

Passions: astronomy, engineering, mathematics, philosophy.

Patron: King Hiero II.

Death: murdered by Romans in 212 BC.

GROUND-BREAKING ACHIEVEMENTS

Pi: Archimedes was the **first known mathematician to estimate pi**—the ratio of a circle's circumference to its diameter.

Archimedes's principle: a floating object displaces its own weight in fluid.

Archimedean screw: Archimedes **designed a revolving screw-shaped blade within a cylinder that would pump water from one level to another**. Versions of it are still in use today.

Death ray: Archimedes is reputed to have **designed deadly mechanisms** using mirrors to focus sunlight onto enemy ships, setting them on fire.

The claw: to defend sea ports, Archimedes **designed a metal grappling hook** suspended from a crane-like arm. The claw would lift one end of an attacking ship out of the water, tipping and sinking the vessel.

Theorems: Archimedes worked out **how to calculate the area of a circle, the surface area and volume of a sphere**, and **the area under a parabola**.

ARCHIMEDES' LEGACY

The lunar mountain range **Montes Archimedes** is named after him.

The **Fields Medal**, awarded for outstanding contributions to mathematics, carries a portrait of Archimedes.

"Eureka!" is the **state motto of California**, coined at the start of the Gold Rush.

Galileo revered Archimedes and stated that he was **"superhuman."**

ERATOSTHENES OF CYRENE

Eratosthenes defined an early algorithm for finding prime numbers, discovered leap years, and was the first recorded person to estimate the Earth's circumference.

LIFE AND TIMES

Born: c.276 BC in **Cyrene**, then a Greek city, now the Libyan town of Shahhat.

Education: studied **philosophy** and **mathematics** in Cyrene and **Athens**.

Passions and writings: **geography**, **mathematics**, **astronomy**, **history**, **comedy**, and **poetry**.

Key role: director at the **Library of Alexandria**, the most prestigious center of learning in the ancient world.

Death: c.195 BC, Eratosthenes is believed to have **gone blind** and **committed suicide by starving himself to death**.

FAMED WORK

Eratosthenes is believed to be the **first scientist to estimate Earth's circumference**. He got a result of **27,400 miles**, close to the **modern value of 24,800 miles**.

He also devised a **calendar** with a 365-day year and an extra day every fourth year.

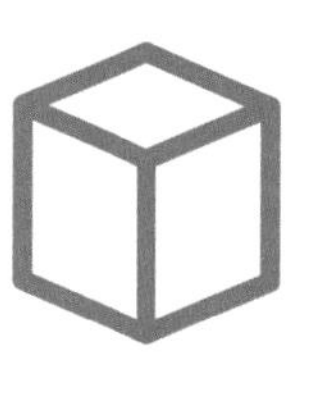

He worked out how, given a **cube**, to calculate the size of a new cube so that its **volume is doubled**.

THE SIEVE OF ERATOSTHENES

1	2	3	4	5	6	7	8	9	10
11	12	13	14	15	16	17	18	19	20
21	22	23	24	25	26	27	28	29	30
31	32	33	34	35	36	37	38	39	40
41	42	43	44	45	46	47	48	49	50
51	52	53	54	55	56	57	58	59	60
61	62	63	64	65	66	67	68	69	70
71	72	73	74	75	76	77	78	79	80
81	82	83	84	85	86	87	88	89	90
91	92	93	94	95	96	97	98	99	100

The ancient Greeks understood **prime numbers**—those that cannot be made by multiplying other whole numbers.

By creating a "**numerical sieve**," Eratosthenes invented a way of finding prime numbers systematically through a **process of elimination**.

First you would remove 1, which is not a prime number, then every number divisible by 2, then those divisible by 3, 5, 7, and 9. Shown above is the result of doing this for all the numbers up to 100.

Eratosthenes had created a **quick and simple mathematical tool**. It was an early "**algorithm**."

CONIC SECTIONS

Conic sections are a class of curves formed by taking slices through a cone. Understanding them was an important development in geometry, with many real-world applications.

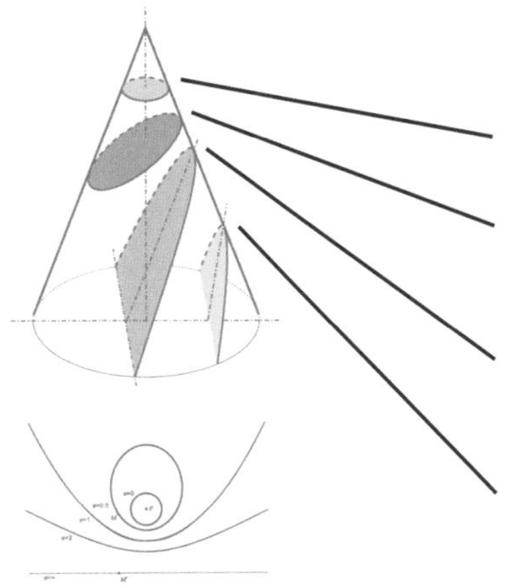

ANATOMY OF A CONIC

A horizontal slice through the cone makes a simple circle.

Tilt the slice **away from horizontal** and the circle deforms into an **oval shape** known as an **ellipse**.

Keep tilting until the slice is parallel with the side of the cone and the figure becomes an **open curve** called a **parabola**.

Tilt further still and the figure becomes an even wider sort of open curve that mathematicians refer to as a **hyperbola**.

KILLER APPS

Conic sections have **applications in the design of optical systems and satellite dishes**. They also describe the **paths of objects moving in electric and gravitational fields**. A ball thrown up from the surface of the Earth follows a **parabola**. But give it enough of a kick and it'll reach **orbit**, moving on a **circular or elliptical trajectory**.

HISTORY

$y=x^2$

The first known study of conic sections was in the fourth century BC by the **Greek mathematician Menaechmus**. Later work was done by **Euclid** and **Apollonius**.

In the year AD 1000, a device for drawing conic sections was designed by the **Arabic scholar Al-Quhi**.

Although conic sections were initially studied from a geometric point of view, in the seventeenth century **John Wallis** was able to describe them using **mathematical equations**.

COMBINATORICS

Combinatorics is a branch of math dealing with the possible combinations and arrangements of objects, for example balls drawn at random from a bag.

FACTORIALS

Given an integer n, the **factorial function**, denoted n!, is a useful device when it comes to combinatorics. It's equal to n × (n – 1) × (n – 2) × ... × 1.

Table of Factorials 1!–30!	
01!	1
02!	2
03!	6
04!	24
05!	120
06!	720
07!	5040
08!	40320
09!	362880
10!	3628800
11!	39916800
12!	479001600
13!	6227020800
14!	87178291200
15!	1307674368000
16!	20922789888000
17!	355687428096000
18!	6402373705728000
19!	121645100408832000
20!	2432902008176640000
21!	51090942171709440000
22!	1124000727777607680000
23!	25852016738884976640000
24!	620448401733239439360000
25!	15511210043330985984000000
26!	403291461126605635584000000
27!	10888869450418352160768000000
28!	304888344611713860501504000000
29!	8841761993739701954543616000000
30!	265252859812191058636308480000000

PERMUTATIONS

Armed with the factorial we can calculate permutations—**the number of ways of arranging a group of objects**. For instance, if you have the numbers 1 to 3, there are six possible arrangements: 123, 132, 213, 231, 312, and 321. In general, **given n items, there will be n! possible permutations**. If you aren't using all n, but want to draw just k items from a possible n, then the formula becomes

$$n!/(n-k)!$$

COMBINATIONS

When you're **not bothered about the order in which the items are arranged, the resulting possibilities are known as combinations**. For example, if you have the numbers 1 to 3 and want to draw two, then there are three possible combinations: 1, 2; 1, 3; and 2, 3. **Order is not important here**; so, for example, 1, 2 = 2, 1.

In general, when drawing k items from n, the number of possible combinations are

$$n!/(k!(n-k)!)$$

This is sometimes written just $\binom{n}{k}$ and referred to as "**n choose k**."

n	k	Permutations	Combinations
2	1	2	2
2	2	2	1
3	1	3	3
3	2	6	3
3	3	6	1
4	1	4	4
4	2	12	6
4	3	24	4
4	4	24	1
5	1	5	5
5	2	20	10
5	3	60	10
5	4	120	5
5	5	120	1

PROBABILITY

Combinatorics are often used in probability theory for calculating the number of ways a particular outcome can happen. For example, if a bag contains three red balls and three black balls, and you're drawing out two at random, then the probability of them both being red is 3 choose 2, divided by 6 choose 2, which is 0.2. It'll happen 20 percent of the time.

BINARY NUMBERS

Binary number systems are the fundamental core of information technology. They evolved in ancient India through the work of mathematician Acharya Pingala.

WHAT IS A BINARY NUMBER?

Whereas our **decimal, base 10, number system** is built from the ten digits 0–9, binary numbers are expressed using **base 2**, which uses only two numerals, typically 0 and 1.

HOW TO FIND A BINARY NUMBER

Binary numbers can be thought of as adding up powers of two.

For example, the number 122 can be written in binary as 01111010; 0 × 128 + 1 × 64 + 1 × 32 + 1 × 16 + 1 × 8 + 0 × 4 + 1 × 2 + 0 × 1. (See table, above right).

Power of 2	2^7	2^6	2^5	2^4	2^3	2^2	2^1	2^0
Decimal	128	64	32	16	8	4	2	1
Binary	0	1	1	1	1	0	1	0

The binary digits act like switches, turning off (0) or on (1) each power of two.

FIRST TEN BINARY NUMBERS

0	1	2	3	4	5	6	7	8	9
0	1	10	11	100	101	110	111	1000	1001

BINARY FACTS

Binary numbers are used in **computing**, where 1 or 0 can be easily **encoded** in the state of an **on-off switch**.

Another common number system is **base 60**, or **sexagesimal**. It was developed by the ancient **Sumerians** over 4,000 years ago and is still used for measuring the passage of time.

Binary numbers were developed into their current form in the seventeenth century by **Gottfried Leibniz**.

PINGALA

Early binary was developed by the scholar **Acharya Pingala**.

Pingala was an **Indian mathematician** who lived more than 2,000 years ago.

He is reputed to be the **first person to use "zero,"** which he called "Śūnya," from **Sanskrit**.

Pingala studied **patterns within poetry**, which he found he could write as **"syllables,"** using **a kind of binary notation**.

ALGEBRA

Algebra is the name we learn in school for the technique of switching numbers with symbols to solve equations. It's also the starting point for many advanced areas of mathematics.

1800 BC Babylonian tablet shows solution of a quadratic elliptic equation

300 BC Euclid's *Elements* includes sections on solving quadratic equations

AD 499 Indian mathematician Aryabhata derives solutions to linear equations

820 Al-Khwarizmi's book *Al-kitāb al-mukhtaṣar fī ḥisāb al-ğabr wa'l-muqābala* is completed

1070 Omar Khayyam investigates cubic equations

1832 Evariste Galois carries out early work on abstract algebra

THE ESSENCE OF ALGEBRA

Algebra enabled scholars to **pose problems mathematically**—and then **solve them**.
For example, consider the equation $2x - 4 = 0$.

This is called a **"linear" equation**, because it involves no powers of x bigger than 1.

From this, you can deduce the value of x: just add 4 to both sides and then divide by 2, and you're left with $x = 2$.

SIMULTANEOUS EQUATIONS

Algebra also makes it possible to solve equations with more than one unknown quantity. This is possible as long as there is one equation for each unknown quantity. So, if we have

$$2x - y = 5$$
$$x + y = 1,$$

we can **add these equations together** to get $3x = 6$, or $x = 2$. And then, **substituting this back into either equation**, reveals that $y = -1$.

NAMING NAMES

The name "algebra" evolved from the Arabic "***al jabr***" (meaning "**the union of broken parts**"), taken from the title of a book on algebraic techniques by the Islamic mathematician **Muhammad ibn Musa al-Khwarizmi**.

HIGHER ALGEBRA

Modern algebra **has moved past the solving of equations**, extending algebraic techniques to **geometry**, **logic**, **number theory**, and more **abstract mathematical concepts**.

TRIGONOMETRY

From surveying to video game design to physics and engineering, trigonometry is a powerful tool for calculating the sizes and proportions of triangles.

WHAT IS IT?

Trigonometry is **a branch of math that relates the ratios of the side lengths of a right-angled triangle to its angles**. For example, if one angle in the triangle is 30°, then the triangle must be the same basic shape as any other right-angled triangle with a 30° angle—meaning that the side lengths must all be in **the same proportions to one another**.

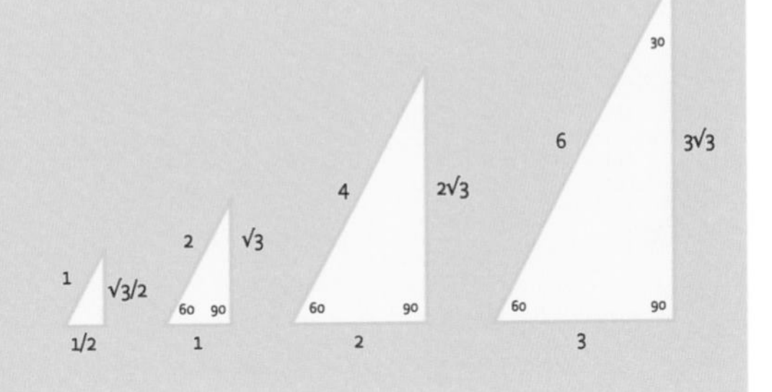

TRIGONOMETRIC FUNCTIONS

There are **three principal trigonometric functions**, known as **sine**, **cosine**, and **tangent**—usually abbreviated to just **sin**, **cos**, and **tan**.

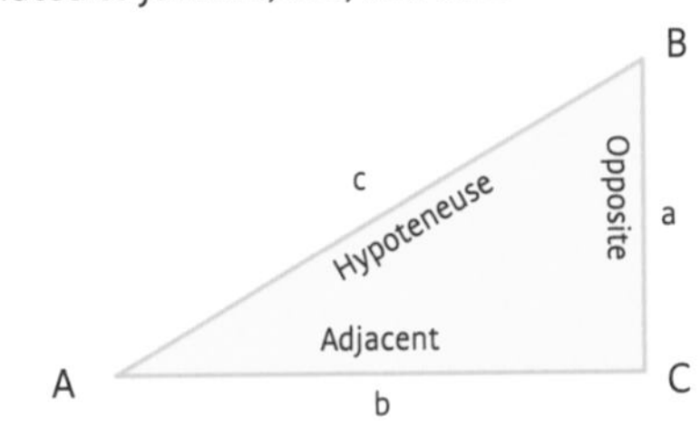

They are defined in terms of side lengths as:

sin A = Opposite/Hypotenuse
cos A = Adjacent/Hypotenuse
tan A = Opposite/Adjacent

Students sometimes like to remember these using the mnemonic SOH-CAH-TOA.

VALUES

As with logarithms, there **was a time when trigonometric functions were laboriously looked up on tables**. Nowadays, numerous **pocket calculators** and **phone apps** can evaluate the functions at the touch of a button. However, some commonly encountered values are:

Angles in degrees	**0°**	**30°**	**45°**	**60°**	**90°**
sin	0	1/2	√2/2	√3/2	1
cos	1	√3/2	√2/2	1/2	0
tan	0	√3/3	1	√3	Not defined

3rd millennium BC Sumerian astronomers develop the study of angles

140 BC Hipparchus gives the first tabulated values of the sine function

1543 Astronomer Nicolaus Copernicus describes trigonometry in his book *De Revolutionibus Orbium Coelestium*

1595 German scholar Bartholomaeus Pitiscus coins the term "trigonometry"

ZERO

For thousands of years, mankind managed without the number zero, but these days it's hard to imagine a world without it.

FROM ZERO TO HERO

Zero is **fundamental** to mathematics and **allowed it to advance enormously**. However, **even the ancient Greeks**, with their progressive ideas in mathematics, **lacked a way of dealing with it**.

It was the Indian scholar **Brahmagupta** who understood that zero was **a number in itself** and had applications in mathematics.

In the twelfth century, the Italian mathematician **Fibonacci introduced zero alongside Hindu-Arabic numerals to Europe**. He was taking a risk, however. The **Church distrusted Arabic numerals** and **regarded nothingness as satanic**.

Zero was banned in Florence in 1299, but merchants found it useful and continued to whisper its Arabic name, "*sifr*." Now zero is **as important as one to nine in our number system**.

WHY ZERO COUNTS

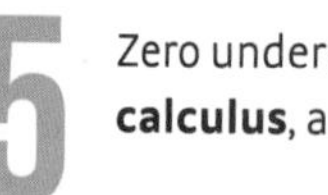

5 Zero underpins **number systems**, **algebra**, **calculus**, and **geometry**.

4 It has a **close relationship with infinity**, as any number divided by zero is infinite.

3 It **gives precision down to microscopic figures**, and **permits huge numbers**.

2 **It separates positive numbers from negatives.**

1 Zero is the **placeholder in our decimal system**.

0 **Mathematics** relies on zero. It **could not function without it**.

3000 BC Babylonians are believed to have used a symbol of two slanted wedges to represent zero

900–300 BC The Sanskrit Bakhshali manuscript contains the oldest known written zero

AD 628 Brahmagupta treats zero as its own number and states rules for its use

1202 Fibonacci introduces "0" to the Hindu-Arabic system of numerals

1545 Gerolamo Cardano publishes his work on negative numbers

PTOLEMY

Regarded as a genius of ancient times, Ptolemy knew that the key to solving astronomical problems lay in mathematics. His work represented a zenith in Greco-Roman science.

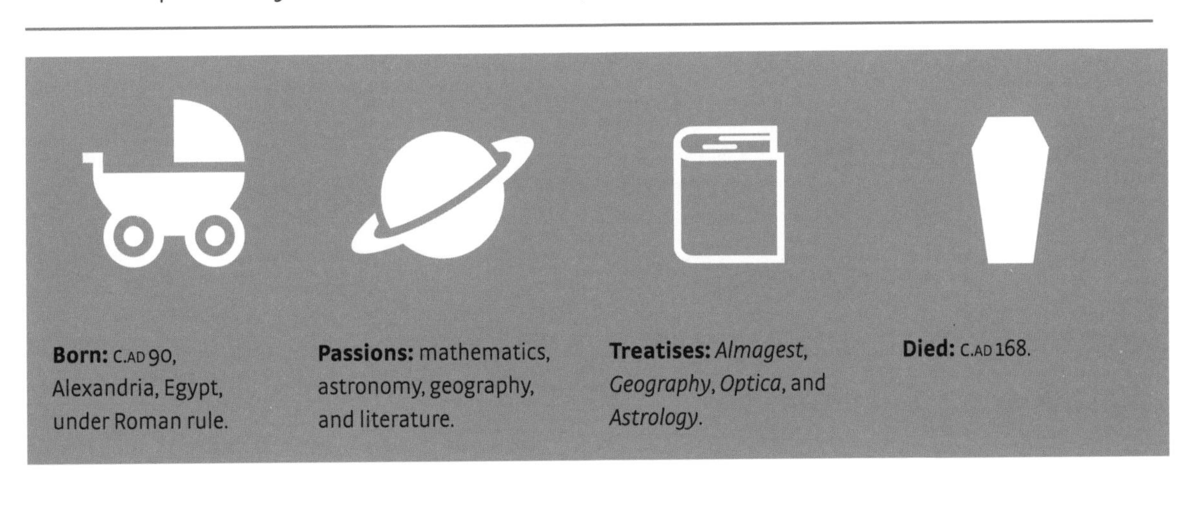

Born: C.AD 90, Alexandria, Egypt, under Roman rule.

Passions: mathematics, astronomy, geography, and literature.

Treatises: *Almagest*, *Geography*, *Optica*, and *Astrology*.

Died: C.AD 168.

MAGNUS OPUS

Ptolemy's seminal work is the ***Almagest***, from the Greek word for "greatest," **a treatise on astronomy** that was completed C.AD 150.

The book **propounds Ptolemy's belief that the motion of the planets, Moon, and Sun could be explained in mathematical terms** and was a key text until the seventeenth century.

Ptolemy claimed to have tabulated **astronomical data** and **geometric models** recorded by **Greek** and **Babylonian** scientists over hundreds of years. These were used to calculate the past and future position of the planets.

The book includes details of **mathematical techniques** such as **spherical trigonometry**.

Almagest also features a **star catalog** of **forty-eight constellations**, building on the work of the Greek astronomer **Hipparchus**.

A GIFTED MAN

Ptolemy's work ***Harmonics*** explored the **mathematics of music** and developed the **Pythagorean theory that musical notes could be turned into mathematical equations**.

In the treatise ***Geography***, Ptolemy charted **longitudes** and **latitudes** in degrees for 8,000 locations on his world map.

In ***Optica***, Ptolemy created an influential treatise on **visual perception**, including an early table on **refraction**, describing how a light ray is bent as it passes **from air to water**.

DIOPHANTUS

An early master of algebra, Diophantus was a Greco-Roman author who wrote the seminal work Arithmetica, *setting out the rules of algebra and inspiring mathematicians for millennia.*

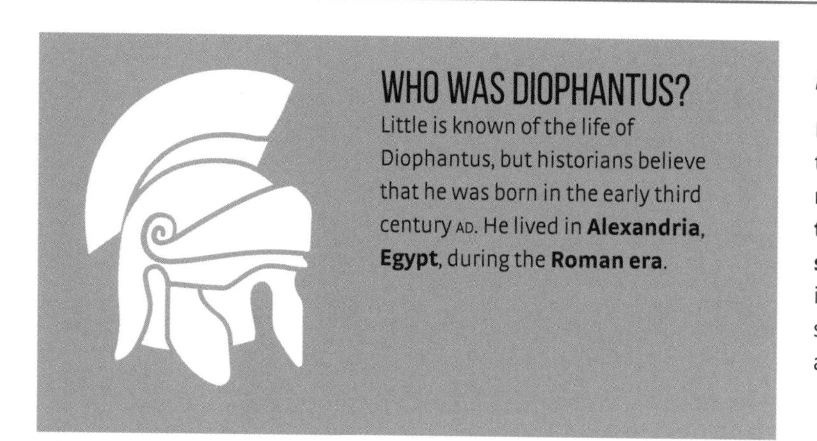

WHO WAS DIOPHANTUS?

Little is known of the life of Diophantus, but historians believe that he was born in the early third century AD. He lived in **Alexandria, Egypt**, during the **Roman era**.

ALGEBRA

Diophantus was the **first known mathematician to use algebraic symbolism**. For instance, he used symbols to represent the unknown as we would use "x."

X

ARITHMETICA

Diophantus is best known for his work *Arithmetica*, which was **the most influential treatise on algebra in early times**. It consisted of thirteen volumes, ten of which are in existence today. Each volume presents a series of problems, with the algebraic methodology required to solve them.

When **Europe** was languishing in the **Dark Ages**, **Arabic scholars** were attempting to solve *Arithmetica*'s most **complex equations**.

HOW OLD?

A puzzle by the fifth-century Greek mathematician **Metrodorus** suggests that Diophantus lived to an age given by solving the **Diophantine equation** $x = x/6 + x/12 + x/7 + 5 + x/2 + 4$.

The solution is $x = 84$.

DIOPHANTINE EQUATIONS

Diophantus pioneered the study of mathematical equations with **purely whole-number solutions**.

$x + 2 = 3$ is Diophantine because its solution is $x = 1$.

But $2x - 1 = 0$ is not, because it has the solution $x = 1/2$.

FERMAT'S LAST THEOREM

It was while reading *Arithmetica* that **Pierre de Fermat** was inspired to propose his **infamous "last theorem"**:

$$a^n + b^n = c^n$$

DIOPHANTINE EQUATIONS

First studied by the great third-century Greek philosopher Diophantus, the Diophantine equations are statements about the relationships between whole numbers.

WHAT ARE THEY?

A Diophantine equation **involves two or more variables raised to powers**. Nothing too unusual about that, except that Diophantine equations deal exclusively with **whole-number**, or **integer solutions**. So, for example, the equation $x^2 + y^2 - 13 = 0$, with the solution $x = 2, y = 3$, is an example. Whereas $x^2 + y^2 - 1 = 0$ (with no integer solutions) is not.

WHY?

$\sqrt{2}$

Diophantus disliked **irrational numbers**—those that **cannot be expressed as a simple, neat fraction**—and so confined his studies instead to the integers. Diophantine equations thus encapsulate the relationships between the integers, and this has spawned a mathematical field in its own right: **number theory**.

WARING'S PROBLEM

Diophantine equations inspired other results in mathematics. In 1770, French scholar **Joseph-Louis Lagrange** demonstrated that, for any integer k, it's possible to choose integers a, b, c, and d such that $k = a^2 + b^2 + c^2 + d^2$ (remember, zero is an integer).

A related question, known as **Waring's problem** (after eighteenth-century English mathematician **Edward Waring**), is **how many terms are needed for powers other than 2**. In 1936, a formula was derived that has since been **tested for all powers up to 471,600,000**. The first few are...

Power	1	2	3	4	5	6	7	8	9	10
Number of terms	1	4	9	19	37	73	143	279	548	1079

10

HILBERT'S TENTH PROBLEM

The so-called **"tenth problem"** of German mathematician **David Hilbert** was to find an **algorithm** to determine whether a given Diophantine-like equation had integer solutions and was thus truly Diophantine. **In 1970, it was shown that no such algorithm exists.**

TESSELLATIONS

These beautiful and strangely hypnotic patterns have been used for millennia in decorative tiling. They are an early example of how mathematics and geometry were applied to daily life.

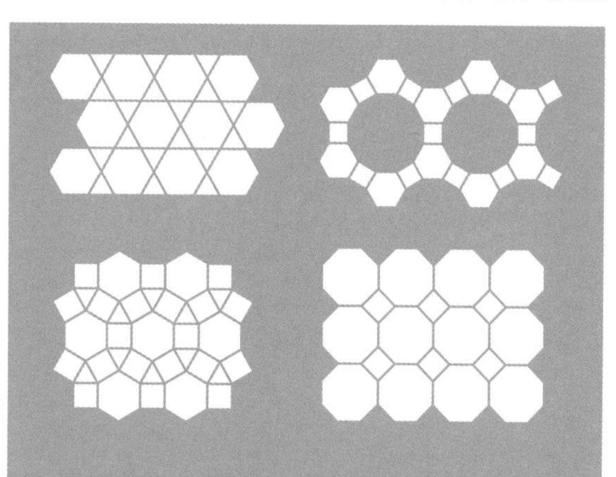

WHAT IS A TESSELLATION?

A tessellation is the **tiling of a flat surface** so that there are **no gaps or overlaps**. When using **polygons**, at each **vertex** (the points formed by the intersection of multiple tiles), the pattern must be the same. The **sum of the angles** at this point will be **360 degrees**.

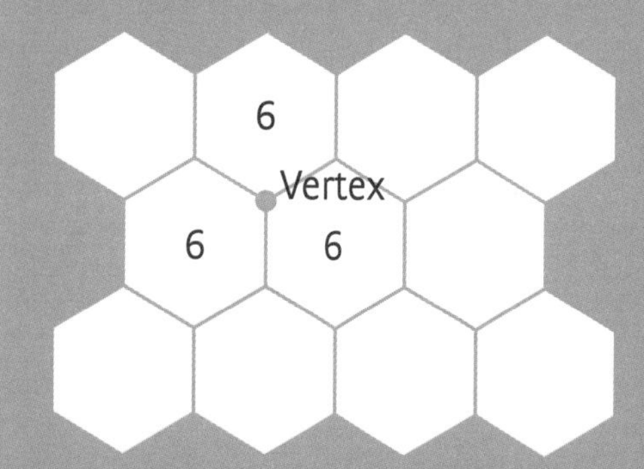

Mosaics, wallpaper, and **floor tiles** often display tessellation. It is also **used in manufacturing to reduce waste**, for instance when **cutting out sheet metal** for **drinks cans** or **car parts**. A **jigsaw puzzle** is also a perfect example of tessellation.

SCHLÄFLI SYMBOL

Two-dimensional tessellations consisting of a single type of regular **polygon** can be specified using the **Schläfli symbol**, two numbers in brackets separated by a comma. The first gives the number of sides each polygon has, while the second gives the number of polygons surrounding each vertex. For example, **a pattern of interlocking regular hexagons** is (6, 3).

EXAMPLES OF TESSELLATION

Monohedral—consisting of one shape. These are often polygons, but they don't have to be, as the spiral illustrates.

Pythagorean tiling—two different-sized squares are used within the pattern.

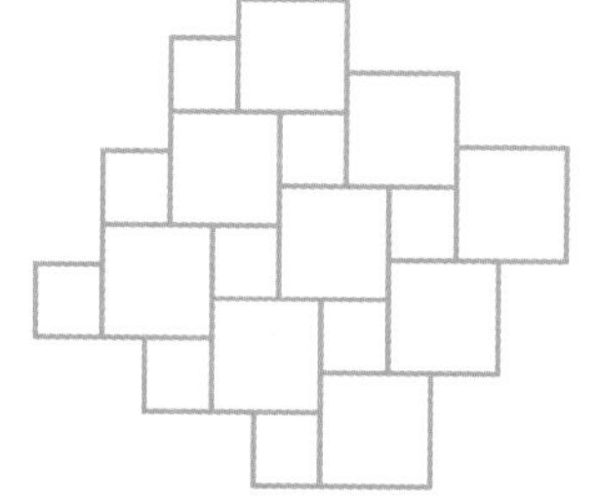

A 3-D tessellation created from interlocking polyhedra, in this case truncated octahedrons.

Nature's marvel. The honeycomb is an example of how tessellations can crop up in nature.

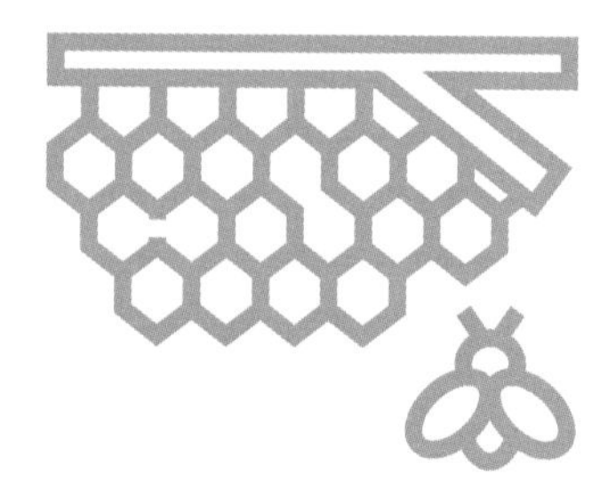

BOETHIUS

Boethius was a talented Roman scholar whose work ensured that the study of mathematics continued in Europe, despite the cultural nadir of the Dark Ages.

Born c.AD 477
in Rome, around 130 years after the fall of the Roman Empire.
Died c.AD 524 in Pavia, northern Italy.

A ROMAN STATESMAN

- Boethius was born into an **aristocratic family**.
- He was orphaned and brought up by the patrician **Symmachus**, who inspired his studies in **literature and philosophy**.
- Boethius became a **consul** in AD 510, a powerful and prestigious position.
- Despite the demands of government, **Boethius began translating the works of Aristotle and Plato from Greek into Latin**.

GREAT INTELLECTUAL

- Boethius compiled a text that **would influence academic study for centuries**.
- His work ***Arithmetica*** was pivotal because it brought together and preserved the work of earlier mathematicians, including **Pythagoras**, **Euclid**, and **Ptolemy**.
- He also completed works on **logic** and **theology**, and developed the connection between **mathematics** and **music**.
- Boethius **ensured that mathematics flourished as a scholarly discipline**.

TRAGIC DOWNFALL

- In AD 522, Boethius was **appointed head of all government and court services in Rome**.
- A year later, he fell out of favor with the king, **Theodoric the Great**, and was **imprisoned under charges of treason**.
- While in prison, Boethius wrote his seminal work ***De Consolatione Philosophiae***.
- In AD 524, Boethius was **executed for treason**.
- His legacy lives on. Boethius's **work helped to bridge the divide between the teachings of the ancient world and those of the modern**.

ARYABHATA

One of the great scholars of India's classical age, Aryabhata helped to develop understanding of mathematics and astronomy in his seminal work Aryabhatiya.

LIFE AND TIMES

Aryabhata was born c.AD 476 in **Kusumapura** (present-day **Patna**), India.

At the age of twenty-three he wrote his classic text *Aryabhatiya*.

Aryabhata had a profound influence on **astronomy**, correctly describing the **physics of eclipses**, and realizing that **planets and the Moon shine because of reflected sunlight**.

He was the **first known mathematician to estimate accurately the value of π**, doing so to the fourth decimal place.

Aryabhata wrote his mathematical and astronomy observations in **verse couplets**.

EARLY GENIUS

Why was *Aryabhatiya* such a ground-breaking work?

It contains a **wealth of mathematical reasoning** on such topics as **geometry**, **solving equations**, and **number series**.

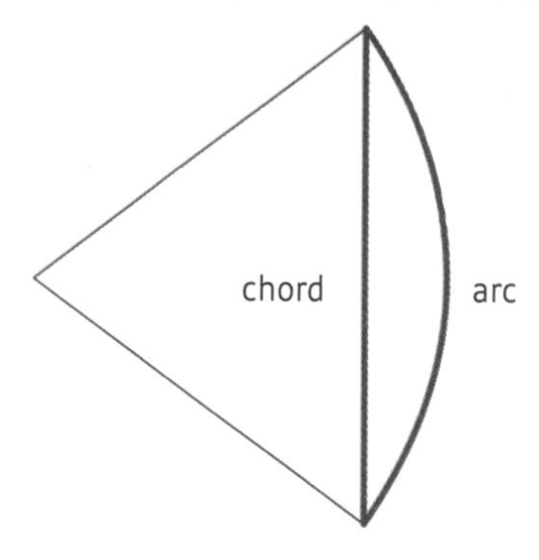

Aryabhata discusses the **trigonometric sine function**, and gives a **table of its values**. He calls the function ***"ardha-jya,"*** meaning **"half-chord,"** which was later translated into the **Latin *"sinus,"*** and the **English "sine."**

.0000000001

Aryabhata makes use of the **"place value" system**, where the value of a digit in a number increases by a **power of ten** for each place that it is **shifted to the left**. For example, 100 is two powers of 10 (i.e., 100) times bigger than 001.

The book **gave elegant formulas for calculating the sum of a series of squared or cubed numbers**. For example:

$$1^2 + 2^2 + \ldots + n^2 = \frac{n(n+1)(2n+1)}{6}$$

A LASTING LEGACY

Aryabhata is thought to have died c.AD 550.

Many **Indian mathematicians** were **inspired to write their own commentaries on *Aryabhatiya***.

In 1975, **India named its first satellite in his honor**.

BRAHMAGUPTA

This Indian mathematician and astronomer was born toward the end of the Indian golden age of learning. Among many discoveries, he identified the crucial importance of zero.

FAST FACTS

Brāhmasphuṭasiddhānta was born in AD 598, in **Rajasthan**, India.

He was the **first mathematician to define rules for handling zero** and to identify it as **a number like any other**.

In AD 628, Brahmagupta explained his theories in the work *Brāhmasphuṭasiddhānta*, which translates as **"correctly established doctrine of Brahma."** Brahma was a Hindu god.

Brahmagupta liked to solve mathematical problems purely for **fun**, and can thus be regarded as one of the **first known recreational mathematicians**.

He died in AD 668, in India.

BRAHMAGUPTA'S FORMULA

Brahmagupta devised a **formula for calculating the area of a cyclic quadrilateral**—that is a four-sided figure that can be inscribed within a circle.

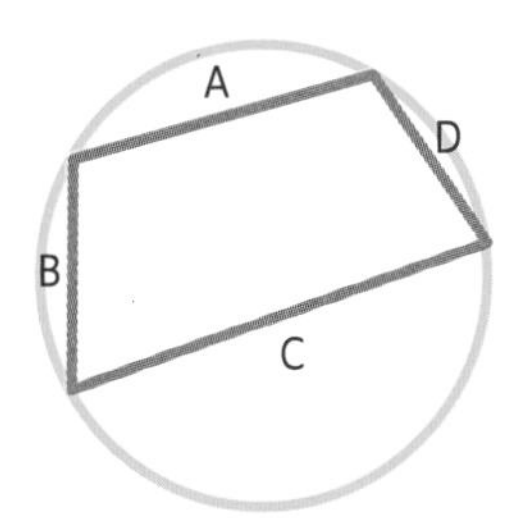

The formula gives the area as

$$K = \sqrt{(s-a)(s-b)(s-c)(s-d)}$$

where a, b, c, and d are the side lengths, and s is one half of their sum.

QUADRATIC EQUATIONS

Brahmagupta gave the **first written statement of a formula for solving quadratic equations**, those of the form $a x^2 + b x + c = 0$. The version of the formula used today gives two viable solutions:

$$x = \frac{-b \pm \sqrt{(b^2 - 4ac)}}{2a}$$

However, Brahmagupta's formula gave just one of these, corresponding to the **positive square root**. For example, $x^2 + 5x + 4 = 0$, has the full solution $x = -2.5 \pm 1.5$, i.e., $x = -1$ or $x = -4$. **Brahamgupta's formula only identified the first of these.**

BRAHMAGUPTA ON ZERO

Brahmagupta used interesting language to illustrate the rules of zero:

A DEBT MINUS ZERO IS A DEBT.

A FORTUNE MINUS ZERO IS A FORTUNE.

ZERO MINUS ZERO IS A ZERO.

MUHAMMAD IBN MUSA AL-KHWARIZMI

While Western Europe languished in the Dark Ages, the East was enjoying the Islamic Golden Age. One of its most notable sons was al-Khwarizmi.

Born around AD 780 in **Khwarazm**, in modern-day **Uzbekistan**.

Al-Khwarizmi studied **science** and **mathematics** at the **House of Wisdom in Baghdad**.

Between AD 813 and 830, he wrote seminal works on **algebra**, **arithmetic**, and **astronomy**.

In around AD 820, he was appointed **astronomer and head of the library at the House of Wisdom**.

Al-Khwarizmi died around AD 850.

THE BIRTH OF "ALGEBRA"

The term **algebra** derives from the **Arabic "*al jabr*,"** which al-Khwarizmi **coined to describe this branch of mathematics**. It means **"re-joining"** or **"completion."** "*Al jabr*" formed part of the title in Arabic of his famous treatise ***The Compendious Book on Calculation by Completion and Balancing*** (C.AD 820). The work established algebra as a **key mathematical discipline**. It was **translated into Latin** in the twelfth century and became an **essential university textbook for 400 years**.

ACHIEVEMENTS

Al-Khwarizmi discovered **clear, systematic rules** for finding **solutions to linear and quadratic equations**.

He showed how **algebraic calculations** could be used to **solve problems** in elements of Persian life at the time, including **trade**, **inheritance**, and **surveying**.

His work introduced the **Hindu-Arabic number system** to the Western world. A translation of his name later evolved into the term **"algorithm."**

He penned work on the **Hebrew calendar**, determining rules for **which day of the week the first day of the month should fall on**.

MATRICES

A matrix is an array of numbers, either square or rectangular. Treating a block of numbers as a single number like this improved the efficiency of algebraic and geometric techniques.

ANATOMY OF A MATRIX

A $m \times n$ matrix has m rows and n columns:

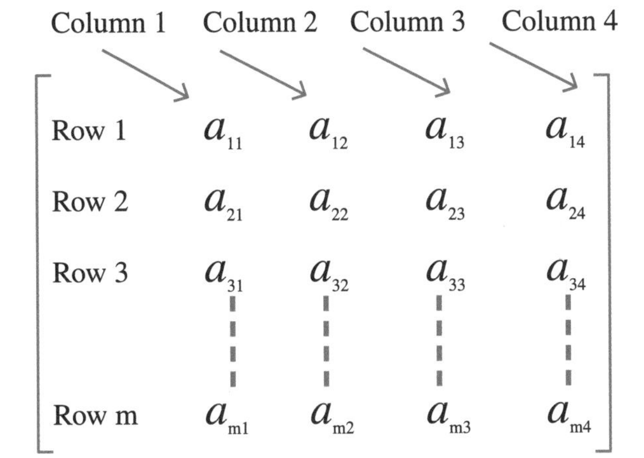

ARITHMETIC

Matrices with the same dimensions can be added and subtracted. Simply add or subtract the corresponding numbers in each matrix.

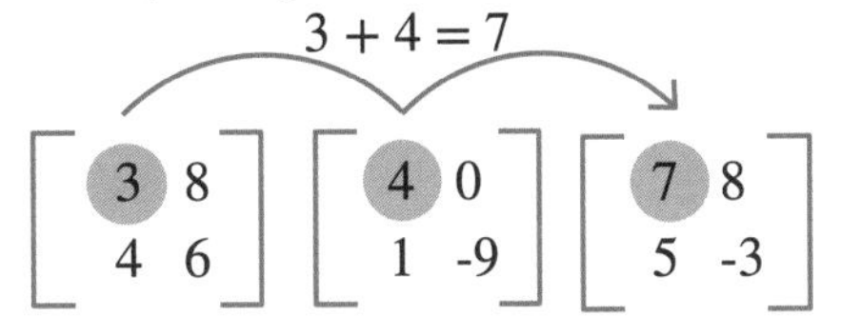

MULTIPLICATION

Multiplying matrices together is slightly more complex. Two matrices, A and B, can only be multiplied if A has the same number of columns as B has rows. It works like this:

$$\begin{bmatrix} A & B \\ C & D \end{bmatrix} \times \begin{bmatrix} E & F \\ G & H \end{bmatrix} = \begin{bmatrix} AE+BG & AF+BH \\ CE+DG & CF+DH \end{bmatrix}$$

Note also that **matrix multiplication does not commute**, so the product $A \times B$ is not the same as $B \times A$.

Matrix division in a sense is also possible, by constructing an **inverse matrix** (written A^{-1}) and **then multiplying by it**. This, though, is only **possible for square matrices**, with **equal numbers of rows and columns**.

ALGEBRA

Matrices are **a way of writing sets of algebraic equations in a neat, condensed form**. For example, **a set of simultaneous equations can be written as a single matrix** equation and then solved simply.

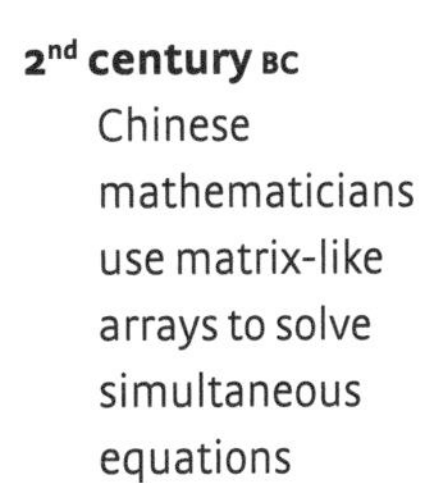

2nd century BC Chinese mathematicians use matrix-like arrays to solve simultaneous equations

1850 English mathematician James Sylvester coins the name "matrix"

1858 Arthur Cayley publishes his book *Memoir on the Theory of Matrices*

1880s Galton and Pearson develop statistical correlation, with matrices linking variables

1925 Physicist Werner Heisenberg uses matrices in his model of quantum theory

OMAR KHAYYAM

Renowned for his poetry as well as astronomy and mathematics, Omar Khayyam made outstanding contributions to the development of algebra.

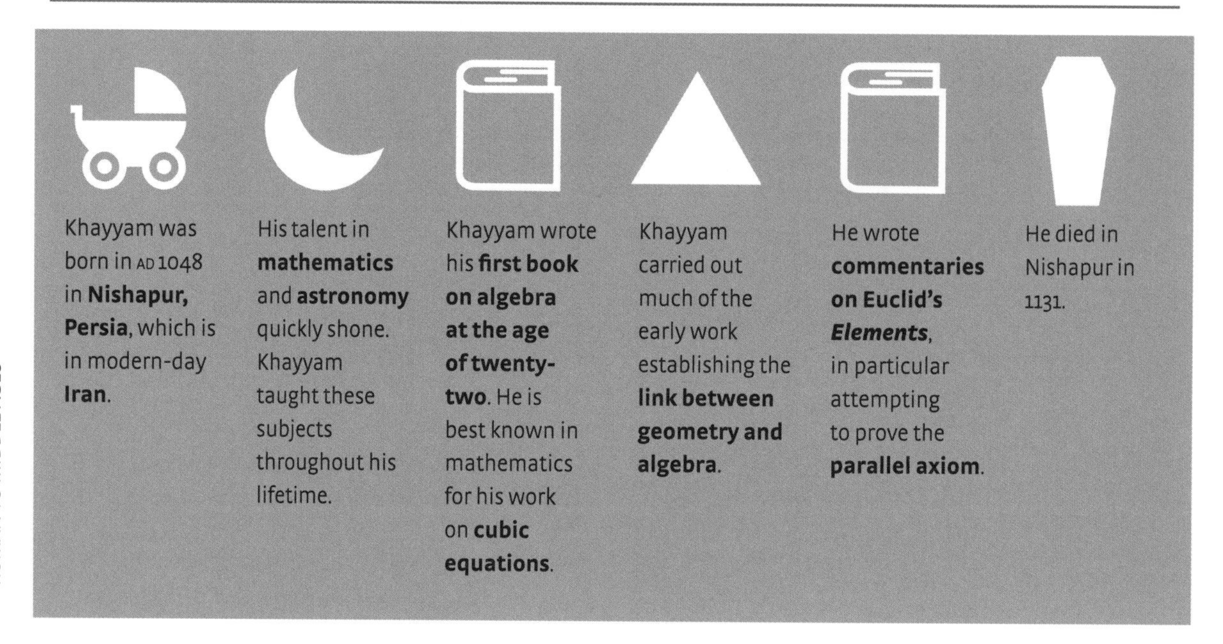

A MASTER OF ALGEBRA

Khayyam's work was crucial to the **development of algebra** as a discipline in modern times.

In his ***Treatise on Demonstration of Problems of Algebra and Balancing***, published c.1070, Khayyam gave extensive examples of how **conic sections such as circles and hyperbolas** can be used to provide the **solutions to cubic equations**.

These are equations involving a cubic (x^3) term, of the form $ax^3 + bx^2 + cx + d = 0$. If b and c are both 0, then solving the equation amounts to taking a simple **cube root**. But, in general, the solution is much **more complicated**.

Khayyam's insight was to **break cubics down** into problems that involved **finding the intersection between two conic sections**. Then by plotting the **conic curves** and finding their **crossing points**, he was able to solve the equation **geometrically**.

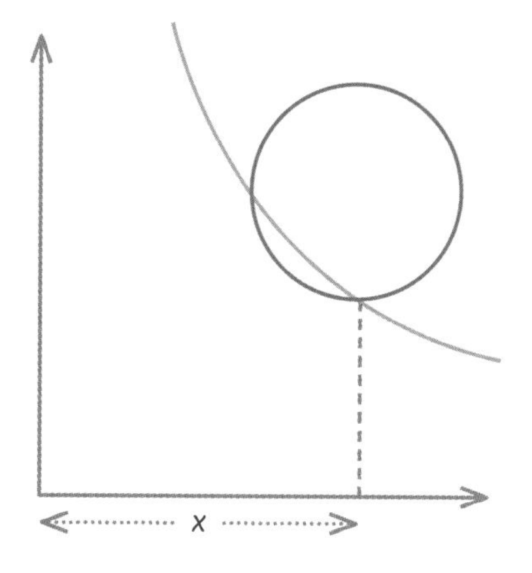

DID YOU KNOW?

The word ***"Khayyam"* means tentmaker**, which may have been his forefathers' profession.

FIBONACCI

Fibonacci changed the world by popularizing the use of Hindu-Arabic numerals in Europe. He is also famous for the eponymous sequence of numbers.

WHO WAS FIBONACCI?

He was born in **Pisa**, Italy, in AD 1170, the **son of a merchant**.

Born **Leonardo**, he was later known as ***filius* Bonacci ("son of Bonacci")**, shortened to Fibonacci.

Fibonacci's father, **Guglielmo Bonacci**, ran a trading post in Bugia, **Algiers**.

Fibonacci **traveled the Mediterranean**, learning about **arithmetic systems**.

In AD 1202, his book ***Liber Abaci* (*Book of Calculation*)** was published.

In AD 1225, his master work ***Liber Quadratorum* (*Book of Squares*)** was published.

He died in around AD 1250, in Pisa.

THE GENIUS OF FIBONACCI

Through his travels, Fibonacci encountered various **numerical systems** and he realized that **Hindu-Arabic numbers** (0–9) were far **more flexible than Roman numerals**, and could be used in **arithmetic**, **trade**, **currency exchange**, **measures**, and **interest**. He summarized his findings in his book *Liber Abaci*, which received acclaim across Europe, where few people had known about Hindu-Arabic numerals.

FIBONACCI SEQUENCE

Try this brainteaser from *Liber Abaci*. "A certain man put a pair of rabbits in a place surrounded on all sides by a wall. How many pairs of rabbits can be produced from that pair in a year if every month each pair begets a new pair, which from the second month on becomes productive?"

The solution is:

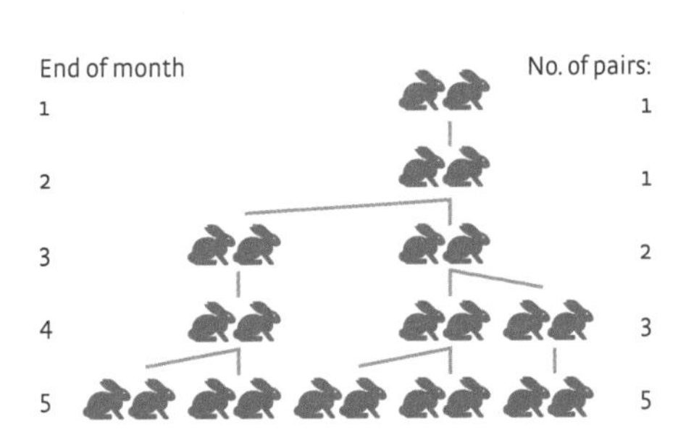

The numbers of rabbits produced forms a **Fibonacci sequence**—a list where **each number is the sum of the two that precede it**. Fibonacci did not name the sequence after himself. It was the nineteenth-century French mathematician **Édouard Lucas** who coined the term. The first few numbers are: 1, 1, 2, 3, 5, 8, 13, 21, 34, 55, 89, 144 ...

The number of rabbit pairs after a year is then the twelfth number in Fibonacci's sequence, namely 144.

QUARTIC EQUATIONS

Solving quadratic equations is hard enough. But beyond these are the cubics and quartics, complex equations that demand some equally complex formulas to solve them.

AD 628 Indian mathematician Brahmagupta gives the first statement of the quadratic formula

1070 Persian scholar Omar Khayyam finds a way to solve cubic equations geometrically

Early 16th century Scipione del Ferro develops a formula to solve cubic equations

1540 Italian mathematician Lodovico Ferrari discovers the solution to quartic equations

1824 The Abel–Ruffini theorem is proven—stating that there are no algebraic solutions to equations of order 5 and higher

THE CUBIC FORMULA

A cubic equation involves integer **powers up to x^3** and takes the general form

$$x^3 + ax^2 + bx + c = 0$$

Solutions can be found using a formula. First define two new variables:

$$q = -\frac{a^3}{27} + \frac{ab}{6} - \frac{c}{2}$$

$$p = q^2 + \left(\frac{b}{3} - \frac{a^2}{9}\right)^3$$

Then, in terms of these quantities, the solutions to the original equation are

$$x = \sqrt[3]{q + \sqrt{p}} + \sqrt[3]{q - \sqrt{p}} - \frac{a}{3}$$

and

$$x = \left(\frac{-1 \pm \sqrt{3}i}{2}\right)\sqrt[3]{q + \sqrt{p}} + \left(\frac{-1 \pm \sqrt{3}i}{2}\right)\sqrt[3]{q - \sqrt{p}} - \frac{a}{3}$$

where *i* is the base imaginary number, equal to $\sqrt{-1}$.

THE QUARTIC FORMULA

The quartic formula is **even more complicated**.

Quartic equations are those involving **integer powers up to and including x^4**. They take the general form

$$x^4 + ax^3 + bx^2 + cx + d = 0.$$

To solve an equation like this, the first step is to define new quantities:

$$e = ac - 4d$$
$$f = 4bd - c^2 - a^2d$$

and then use the cubic formula above to find the solutions of

$$y^3 - by^2 + ey + f = 0.$$

Take the real-valued solution (i.e., with no imaginary part), and set this equal to *y*. Then define

$$g = \sqrt{(a^2 - 4b + y)}$$
$$h = \sqrt{(y^2 - 4d)}$$

And then, finally, the solutions to the original quartic equation (there are four of them) are given by solving the two quadratic equations:

$$x^2 + \frac{1}{2}(a + g)\,x + \frac{1}{2}(y + h) = 0$$

$$x^2 + \frac{1}{2}(a - g)\,x + \frac{1}{2}(y - h) = 0$$

GEROLAMO CARDANO

One of the greatest mathematicians of the Renaissance, Cardano introduced negative numbers into common usage and was one of the first to apply math to gambling.

WHO WAS CARDANO?

He was born illegitimately in 1501, in **Pavia, Italy**. His father **Fazio Cardano** was a friend of **Leonardo da Vinci**.

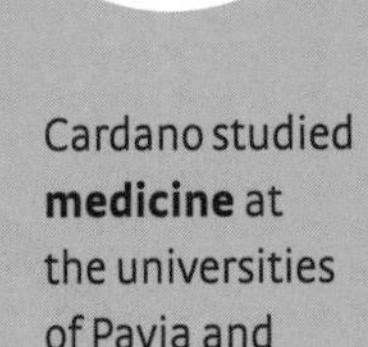

Cardano studied **medicine** at the universities of Pavia and **Padua**.

He was a **prolific author** and wrote works on **mathematics**, **science**, and the **natural world**.

Cardano had **three children**, one of whom was **beheaded for poisoning his wife**.

Cardano was an **obsessive gambler** and played **cards**, **dice**, and **chess** for money.

He died in 1576, in Italy.

THE POWER OF NEGATIVE NUMBERS

A negative number is one that is **less than zero**. The existence of these integers is indisputable in modern times, but **for centuries even great mathematicians regarded them as dubious, false, or downright absurd**.

Cardano published his workings in the masterpiece, ***Ars Magna*** (1545), a forty-chapter treatise on algebra. The work was highly successful and **mathematicians began to accept the legitimacy of negative numbers**.

NEGATIVE NUMBERS IN DAILY LIFE

Depth below sea level.

Sub-zero temperature scales.

Latitudes below the equator.

Bank statements, when they are in the red!

DID YOU KNOW?

Cardano gambled in order to support his income. He wrote a book on probability, ***Liber de Ludo Aleae* (*On Casting the Die*)**.

IMAGINARY NUMBERS

Despite their name, imaginary numbers are as real as any other mathematical entity. Today, they are used in fields including engineering, pure math, and physics.

ROOT OF ALL EVIL

Imagine you're asked to solve the equation $x^2 = -1$. You'd expect to take the square root of both sides to get an expression for x, but in this case that leaves you with $x = \sqrt{(-1)}$. The trouble is that **the square root of a negative number doesn't exist—square any number and you get a positive quantity**.

THE IMAGINARY UNIT

Rather than dismissing such cases as meaningless curiosities, mathematicians created the entity i to denote $\sqrt{(-1)}$, and then treated it as a number like any other. It has the properties

i^2	-1
i^3	-i
i^4	1
i^5	i
i^6	-1

Using i, you can write the square root of any negative number. For example, $\sqrt{(-16)} = \sqrt{16} \times \sqrt{(-1)} = 4i$.

COMPLEX NUMBERS

In modern math, numbers can have both real and imaginary parts. These are called complex numbers. They are typically represented on a 2-D plane, with the vertical axis corresponding to the imaginary part and the real part plotted horizontally. So the complex number 4 + 3i, might look like:

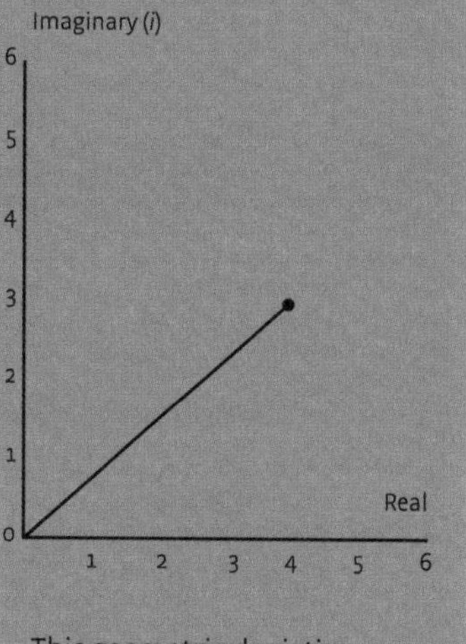

This geometric depiction is known as an **Argand diagram**, after **Swiss mathematician Jean-Robert Argand**.

1572 Italian mathematician Rafael Bombelli lays out the rules governing imaginary numbers

1637 Descartes writes about imaginary numbers in his book *La Geométrié*

1748 Euler shows how complex numbers can be used to represent waves

1777 Leonhard Euler first uses the symbol i to denote $\sqrt{(-1)}$

1843 William Rowan Hamilton extends 2-D complex numbers to 4-D "quaternions"

i, j, k

KEPLER'S CONJECTURE

What's the most efficient way to stack apples in a box? German mathematician Johannes Kepler had a good idea, but it took over 400 years for his suspicions to be proven correct.

ESSENCE OF THE PROBLEM

Throw a load of spheres at random into a box and they will stack themselves such that the spheres themselves occupy about 0.65 of the volume in the box (while the space between them takes up the remaining 0.35).

However, it is possible to do better. If you stack the spheres carefully into either a face-centered cubic (fcc), or a hexagonal close-packed (hcp) configuration, then they can be made to take up

$$\frac{\pi}{3\sqrt{2}} = 0.740480489...$$

of the volume available in the box.

Kepler's conjecture asserts that this is the best you can do—there is no more efficient way to pack spheres together.

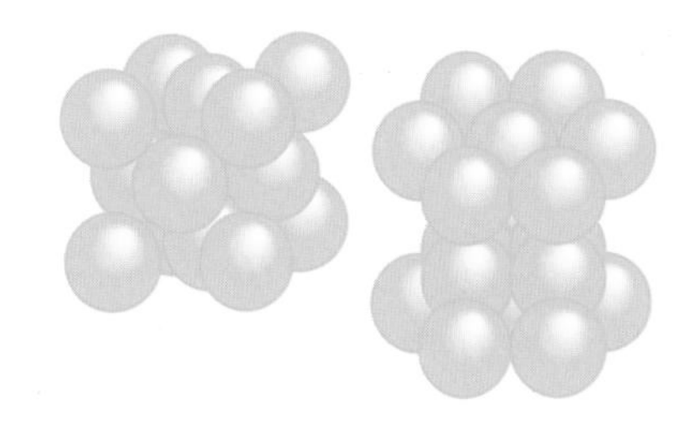

Face-centered cubic (left), and hexagonal close-packed (right) arrangements.

WHO WAS KEPLER?

Born December 27, 1571, in Weil der Stadt, Germany.

Kepler is appointed imperial mathematician to Holy Roman Emperor Rudolf II in 1601.

He publishes the first of his laws of planetary motion in 1609.

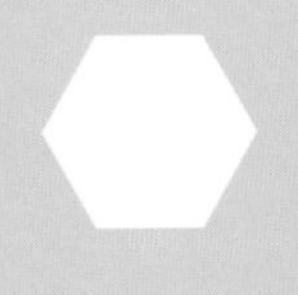

He proposes the Kepler conjecture in 1611 as part of a work on the hexagonal symmetry of snowflakes.

Died November 15, 1630, in Regensburg, Germany.

HOW WAS IT PROVEN?

In August 2014, the conjecture was finally proven by Thomas Hales, of the University of Pittsburgh, working with a team of twenty-one others. It was essentially a "proof by exhaustion," whereby the volume of every possible arrangement of spheres was checked by computer.

JOHN NAPIER

This brilliant Scottish mathematician and physicist discovered logarithms, and developed an early calculating device.

WHO WAS JOHN NAPIER?

Napier was born in **Edinburgh**, in 1550, the son of **Sir Archibald Napier**.

He was educated at the **University of St. Andrews** and in Europe.

Napier lived in **castles** at **Gartness, Stirling,** and **Merchiston, Edinburgh**.

His interests included **mathematics, theology**, and **the occult**.

He died in Edinburgh, in 1617.

THE INVENTION OF LOGARITHMS

As a Scottish land owner, Napier had time on his hands, which he devoted to the study of mathematics. He discovered **logarithms**, which he **first explained and tabulated** in his ground-breaking work ***Mirifici Logarithmorum Canonis Descriptio (Description of the Marvelous Canon of Logarithms, 1614)***.

They express **the power to which a number has to be raised to get another number**. For instance, $10^2 = 100$. Therefore, log "to the base" 10 of 100 = 2.

NAPIER'S BONES

Napier's bones, also known as **Napier's rods**, was **an early mechanical calculating machine** which had multiplication tables embedded into wooden rods. It could even extract square roots.
For example, to multiply 425 by 6, you would take the rods for 4, 2 and 5 and then read off the row corresponding to the number 6:

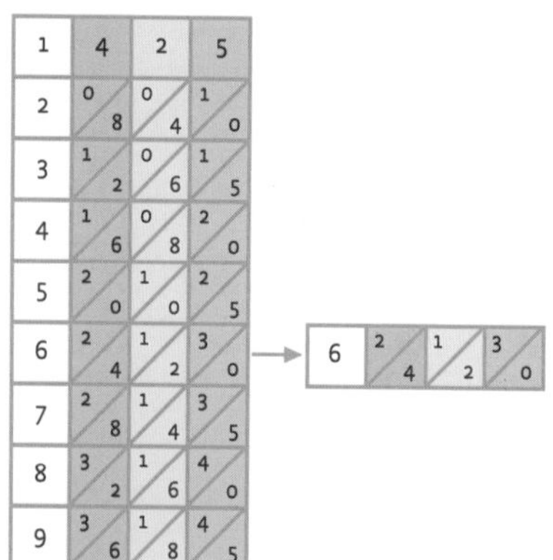

Then, summing the diagonals gives the answer. In this case, 6 × 425 = 2550:

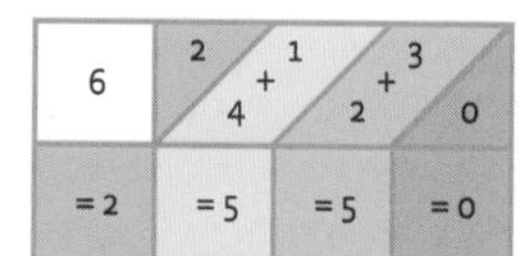

DID YOU KNOW?

The Moon crater Neper is named after Napier (from the Latin form of his name, Ioannes Neper).

Napier used the full stop as a decimal point in his logarithm tables, which brought it into common mathematical usage.

Today, Edinburgh Napier University is named after the great man.

LOGARITHMS

Logarithms are the opposite of raising a number to a power. Their applications in mathematics and science are manifold.

WHAT IS A LOGARITHM?

A power law, given by the equation $y = b^x$, can be turned around and written instead as $x = \log_b y$. In other words, $\log_b$ (which is said as "log to the base b"), tells you the power to which b must be raised in order to get y. They were introduced by Scottish mathematician John Napier in the early seventeenth century.

LOG LAWS

Logarithms obey some very handy rules.

Products:	$\log A + \log B = \log AB$
Ratios:	$\log A - \log B = \log \frac{A}{B}$
Powers:	$\log A^n = n \log A$
Roots:	$\log \sqrt[n]{(n)} = \frac{1}{n} \log A$
Changes of bases:	$\log_b A = \log_n A / \log_n b$

WHY?

Logs are useful when it comes to solving equations for which the exponent (or, the power) is the unknown quantity you're trying to find.

For example, if we are told that $10^x = 100$, solve for x, then we can say that $x = \log_{10} 100$. And consulting a set of log tables (or, more likely, an electronic calculator) reveals that $\log_{10} 100 = 2$.

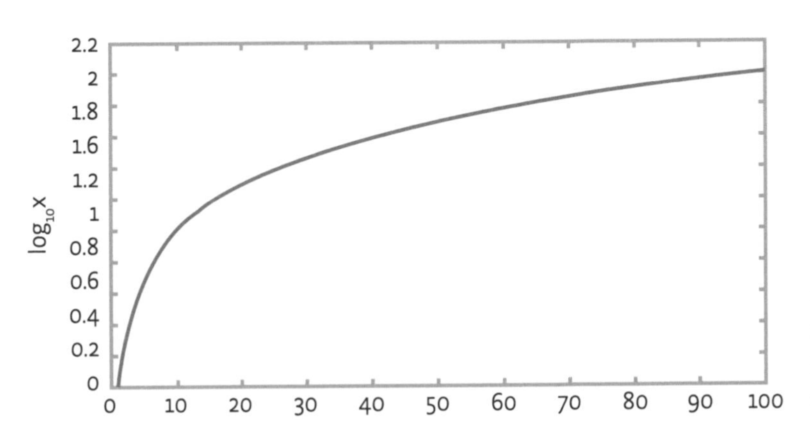

APPLICATIONS

Logs crop up in science as a convenient way of handling very large numbers.

The response of human senses to stimuli is logarithmic.

That's why scales to measure levels of light and sound are often logarithmic too. For example, an increase in sound power of 10 on the decibel scale is ten times more powerful. That is, if the sound increases in power from Po to P then the decibel increase is $10 \log_{10}(P/P_0)$.

Logs also feature in nature. In the "logarithmic spiral," for instance, the radius of the spiral curve scales up by a fixed factor (A), rather than a fixed amount (B), on each turn.

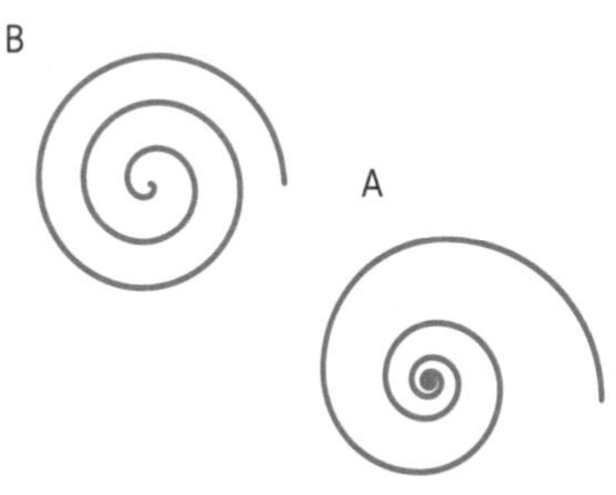

Logarithmic spirals are seen in the shells of the nautilus—a marine mollusc—and the spiral arms of galaxies.

EARLY MODERN

EULER'S NUMBER

Euler's number is a mathematical constant that crops up throughout pure mathematics and forms the basis of so-called "natural logarithms."

WHAT IS IT?

Euler's number is a mathematical constant usually just written as e (after **Leonhard Euler introduced the notation in 1731**), and takes the value 2.71828 ... Like many other mathematical constants, that's an approximation, as e is a **transcendental number** with an **infinite number of digits**.

e^0	1.0000000000000
e^1	2.7182818284590
e^2	7.3890560989304
e^3	20.085536923186
e^4	54.598150033138
e^5	148.41315910255
e^6	403.42879349265
e^7	1,096.6331584282
e^8	2,980.9579870409
e^9	8,103.0839275729
e^{10}	22,026.265794799
e^{11}	59,874.141715175
e^{12}	162,754.79141893
e^{13}	442,413.39200871
e^{14}	1,202,604.2841641
e^{15}	3,269,017.3724702
e^{16}	8,886,110.5205025
e^{17}	24,154,952.753560
e^{18}	65,659,969.137287
e^{19}	178,482,300.96306
e^{20}	485,165,195.40943
e^{21}	1,318,815,734.4822
e^{22}	3,584,912,846.1287
e^{23}	9,744,803,446.2408
e^{24}	26,489,122,129.821
e^{25}	72,004,899,337.323
e^{26}	195,729,609,428.66
e^{27}	532,048,240,601.30
e^{28}	1,446,257,064,290.0
e^{29}	3,931,334,297,139.9
e^{30}	10,686,474,581,513
e^{31}	29,048,849,665,215
e^{32}	78,962,960,182,591

DISCOVERY

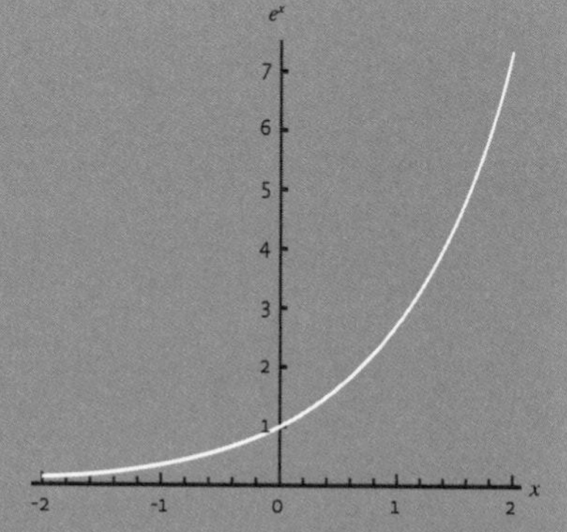

Euler's number was discovered in 1683 by Swiss mathematician **Jacob Bernoulli**, in a study of **compound interest**.

He reasoned that if you put $1 in a bank that pays annual interest of 100 percent once per year, then after a year you have $(1 + 1)^1 = \$2$. If the annual rate is the same but calculated biannually, then after a year you have $(1 + 1/2)^2 = \$2.25$. In general, if you're paid out n times per year then after twelve months you have $(1 + 1/n)^n$. And as n gets very large, this tends toward $2.71828 ..., or e.

NATURAL LOGS

In essence, a natural log is just **like an ordinary logarithm, but to base e**. The natural log of a number, x, is usually written $\log_e x$ or ln x.

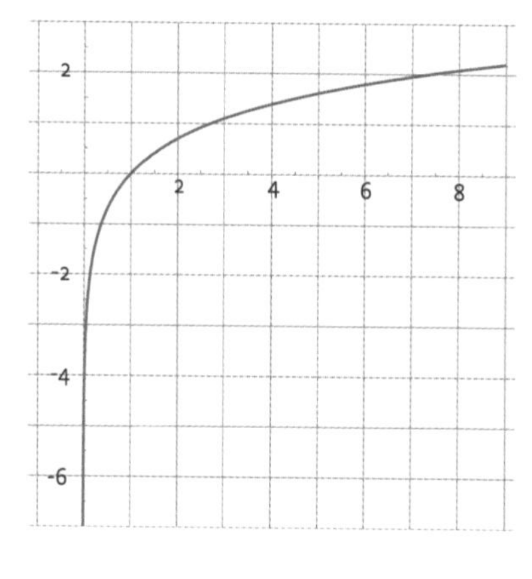

Tables of natural logs were first drawn up by the **English mathematician John Speidell in 1619**.

Natural logs and e also emerge in **calculus**—the branch of mathematics dealing with **how quantities change relative to one another**.

RENÉ DESCARTES

Best known as a philosopher ("I think, therefore I am"), Descartes was also a brilliant mathematician, whose work shaped the course of mathematics in modern times.

WHO WAS DESCARTES?

René Descartes was born in **La Haye en Touraine**, France, in 1596. The village was renamed in his honor.

He studied **law** at the **University of Poitiers**.

In 1618, he joined the **Dutch States Army**.

As a **wealthy man**, Descartes could **devote his life to study**.

In 1650, he died of **pneumonia** in Stockholm.

ANALYTIC GEOMETRY

In 1637 Descartes' key mathematical work *La Géométrie* was published, in which he **pioneered the use of coordinates, allowing geometric curves to be specified algebraically**. This was ground-breaking at the time and **allowed previously impossible geometric problems to be solved**.

INFLUENCE

Descartes contributed to the science of optics, using geometric techniques to demonstrate results in reflection and refraction.

He is also thought to have been a major source of inspiration to the young Isaac Newton.

DID YOU KNOW?

Descartes used letters x, y, and z to express unknown quantities, giving us the system that we recognize today. He expressed knowns as a, b, and c.

He pioneered the use of exponential notation, such as a^3 instead of a × a × a.

Descartes illustrated the fundamental place of algebra in mathematics. Before this, geometry had superseded it in importance.

His work formed the basis of Newton and Leibniz's calculus.

CARTESIAN COORDINATES

Cartesian coordinates enabled geometric shapes to be expressed numerically. Today, these measures are used in navigation, engineering, and space flight.

GEOMETRY TO ALGEBRA

In 1637, **René Descartes**'s book *La Géométrie* was published. In it, Descartes pioneered the use of coordinates. He realized that **a point on a 2-D plane could be described by using its horizontal (x) and vertical (y) locations**. The coordinates of a point are written (x, y).

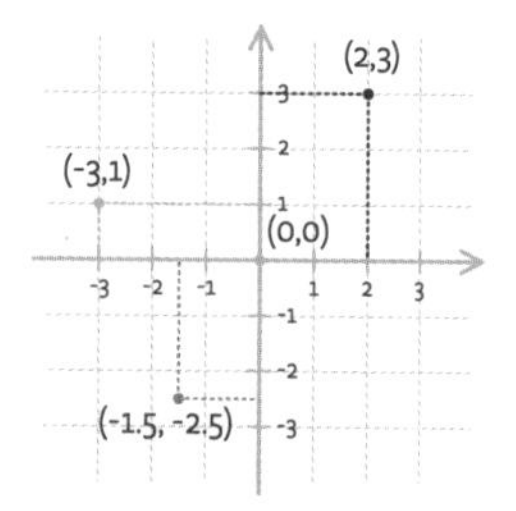

Lines or curves can then be plotted based on **algebraic equations**—a practice known as **analytic geometry**. For example, the equation y = 2x + 3 describes a straight line passing through the points (–2, –1) and (2, 7)...

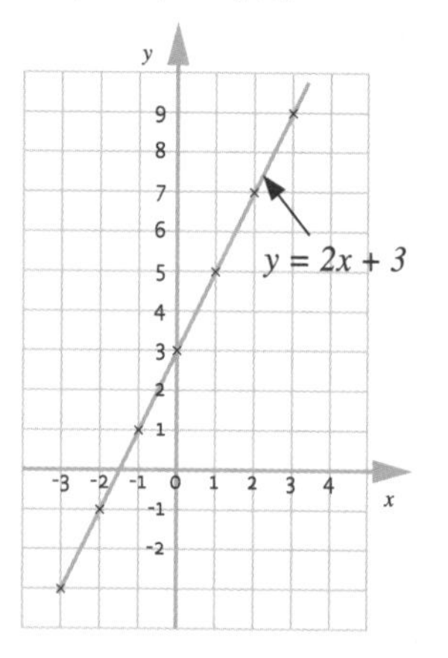

VECTORS

As well as position, Descartes's formalism **allowed quantities to be expressed with both size and direction**. So, for example, the velocity of a ship moving at four knots in the x direction and 3 knots in the y direction can be denoted as (4, 3). This is called a **vector**.

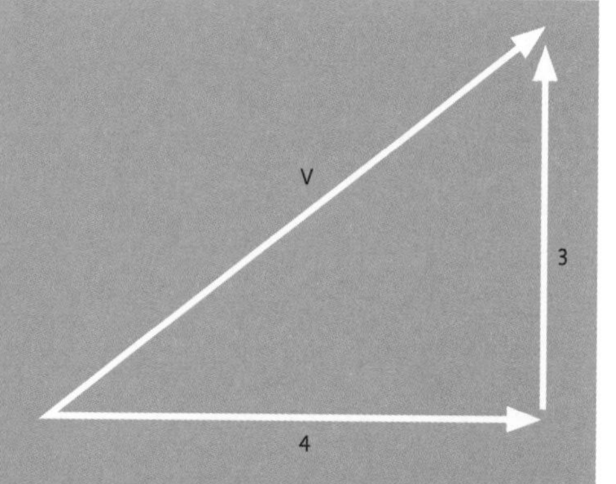

The **Pythagorean theorem** then lets you calculate the overall speed of the ship: $V = \sqrt{(4^2 + 3^2)} = 5$ knots.

DID YOU KNOW?

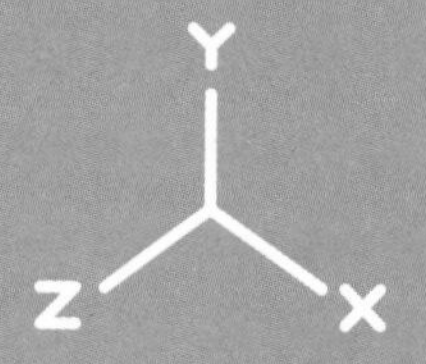

In **3-D space**, Cartesian coordinates are **supplemented with an extra direction**, labeled z.

(r, θ)

Other coordinate systems followed, such as **polar coordinates**, where **position is specified by a distance and an angle**.

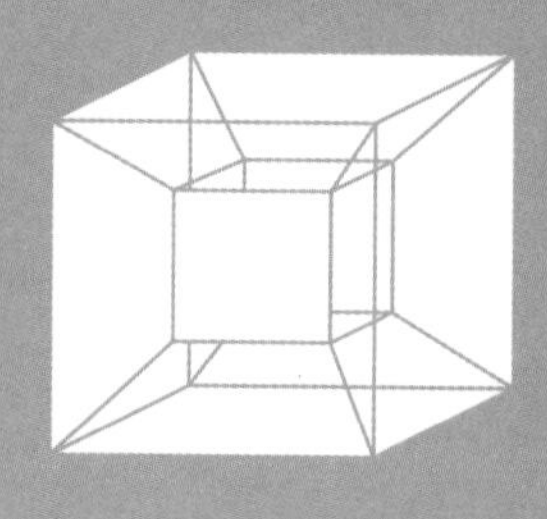

Analytic geometry **allowed spaces to be studied in any number of dimensions**, beyond the three that we can easily "visualize."

PIERRE DE FERMAT

Fermat was a seventeenth-century lawyer and politician who was fascinated with mathematics and science. He conducted pioneering research into calculus, number theory, and the behavior of light.

WHO WAS PIERRE DE FERMAT?

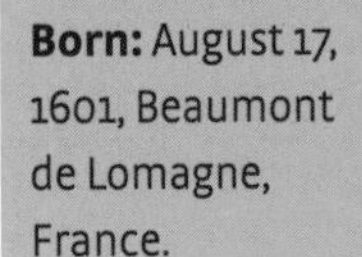

Born: August 17, 1601, Beaumont de Lomagne, France.

Studied: law at the University of Orléans.

Served: local parliament.

Peers: René Descartes and Blaise Pascal.

Interests: geometry, probability theory, calculus, science, literature, and languages.

Died: January 12, 1665, in Castres.

A GENIUS AT WORK

Fermat had an incredible thirst for knowledge. Not only was he a lawyer, he also **spoke six foreign languages**. Mathematics was mostly a **hobby** for him. He communicated his theorems by letter to friends or peers, such as **Blaise Pascal**, who shared his interest in **number theory**. For reasons unknown, Fermat was **reluctant to publish his prolific work**, which limited his influence.

GROUND-BREAKING DISCOVERIES

Analytic geometry: Fermat co-founded (with **Descartes**) this powerful field, which **describes geometrical shapes numerically**. It's used today in **rocket science**, **aviation**, and **engineering**.

Fermat's principle: In 1622, Fermat established the **"principle of least time"**—that a **ray of light traveling between two points in a system of mirrors and lenses will take the quickest path**.

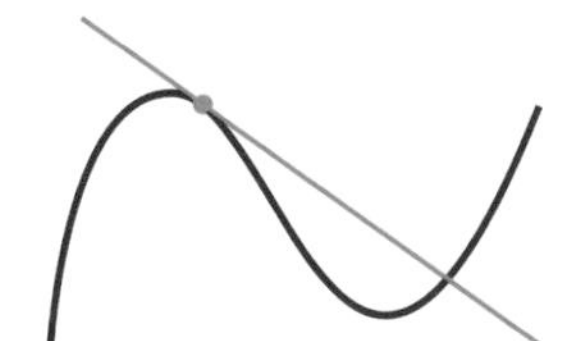

Differential calculus: Fermat developed a method of finding tangents to curves. Differential calculus was further developed by **Sir Isaac Newton** and **Gottfried Leibniz**.

Fermat's last theorem: He conjectured in 1637 that no three positive integers a, b, and c satisfy the equation $a^n + b^n = c^n$ where n is greater than 2. It took until 1994 to prove.

FERMAT'S LAST THEOREM

In the seventeenth century, Pierre de Fermat stated a theorem, but gave no supporting calculations. It took over 350 years to prove him right.

THE THEOREM

The theorem states that the **Diophantine equation**

$$a^n + b^n = c^n$$

cannot be satisfied for any positive integers a, b, and c, when the power n is an integer bigger than 2.

When n = 1 the problem is trivial, and when n = 2 the solutions are just the **Pythagorean triples**.

THE MYSTERY

?

In 1637, French mathematician Fermat scrawled the theorem in the margin of his copy of Diophantus's book *Arithmetica*, along with the claim that he'd found a "truly marvelous" proof—which the margin itself was too small for. After his death in 1665, however, no proof could be found. The claim became known as "Fermat's Last Theorem," and the race was on to prove it.

SPECIAL CASES

Fermat did leave a proof for the case when n = 4, and proofs were later found for other special cases...

n	Author	Date
4	Fermat	Seventeenth century
3	Euler	1770
5	Legendre and Dirichlet (independently)	1825
6	C.F. Kausler	1802
14	Dirichlet	1832
7	Gabriel Lame	1839
10	H. Kapferer	1913

GENERAL SOLUTION

$y^2 = x^3 - 1$ $y^2 = x^3 + 1$ $y^2 = x^3 - 3x + 3$ $y^2 = x^3 - 4x$ $y^2 = x^3 - x$

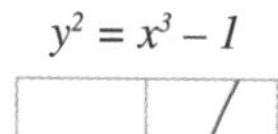

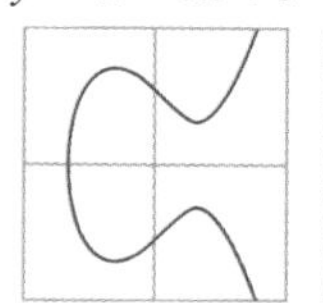

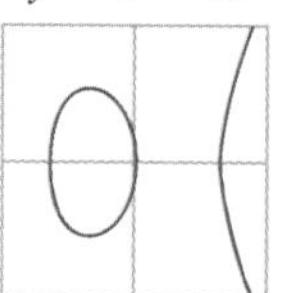

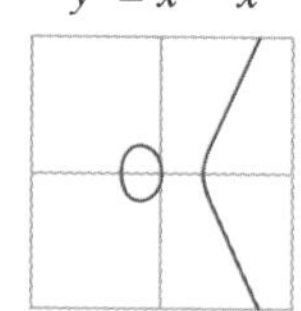

The theorem was ultimately proved for arbitrary n by British mathematician Andrew Wiles, with help from his former student Richard Taylor, in 1994. Wiles used an approach based on the "modularity theorem" for elliptic curves. The proof was published in *Annals of Mathematics* in 1995—358 years after Fermat's original claim.

BLAISE PASCAL

Pascal lived a short life, plagued by illness, but he changed the world with works on probability theory, geometry, and Pascal's triangle, and he invented the world's first working calculator.

- **June 19, 1623** Born in Clermont-Ferrand, France. His father, Étienne, was a respected mathematician and judge of the city's tax court
- **1640** At age sixteen, Pascal publishes his *Essai pour les Coniques* (Essay on conic sections), which included Pascal's theorem
- **1645** After working on fifty prototypes, Pascal presents his calculator to the public
- **1654** Collaborates with Pierre de Fermat on mathematical theory of probabilities in relation to gambling
- **August 19, 1662** At the age of thirty-nine, Pascal dies after suffering what was thought to be meningitis and a stomach ulcer

PASCAL'S CALCULATOR

This device was a **portable wooden box consisting of eight metal wheel dials with numerals from 0 to 9**. These could be **rotated to add and subtract**, giving figures of up to eight digits, which appeared in boxes at the top.

CONTRIBUTIONS TO SCIENCE

PASCAL'S LAW

Pascal invented the **first hydraulic press** and the **syringe**.

He formulated **Pascal's law**—in a fluid at rest in a container, **pressure change** in one part is **transmitted without loss** to the walls of the container.

SI (metric) unit of pressure was later named **the Pascal**.

PASCAL'S THEOREM

Pascal found that if six arbitrary points on **a "conic" (a class of curves, the simplest of which is an ellipse)** are joined by straight lines in an arbitrary order, then the **crossing points of the lines form a straight line**. This is called the **"Pascal line."**

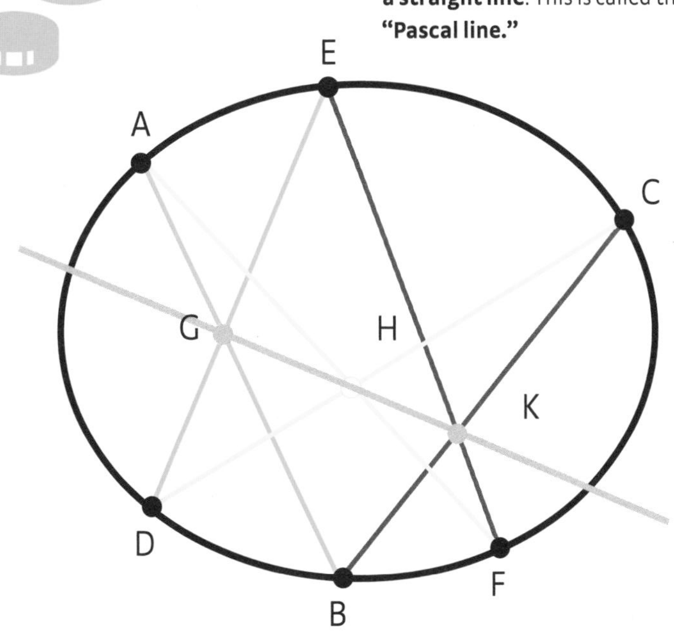

PASCAL'S TRIANGLE

Pascal's triangle is an arrangement of numbers with rich patterns and seemingly magical connections to other branches of mathematics.

MAKE YOUR OWN

To construct, start with a 1 at the apex on the top row. **Entries in the rows below are the sum of the two entries from the row directly above**—with the **edges all taking the value 1**.

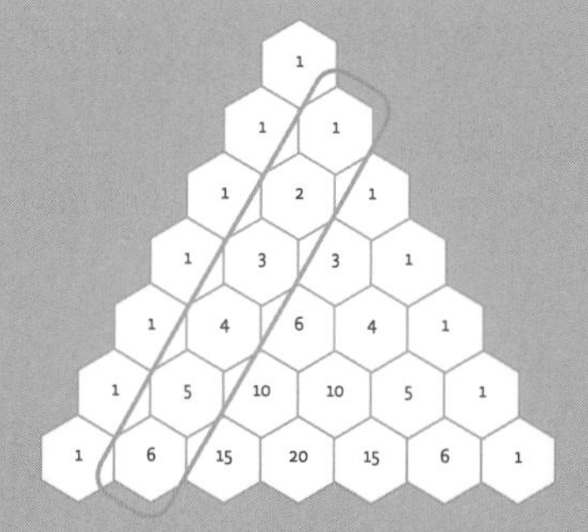

NATURAL NUMBERS

Move inward, parallel to the edge, and the **next sequence gives the natural numbers** (i.e., the positive integers 1, 2, 3, 4, 5).

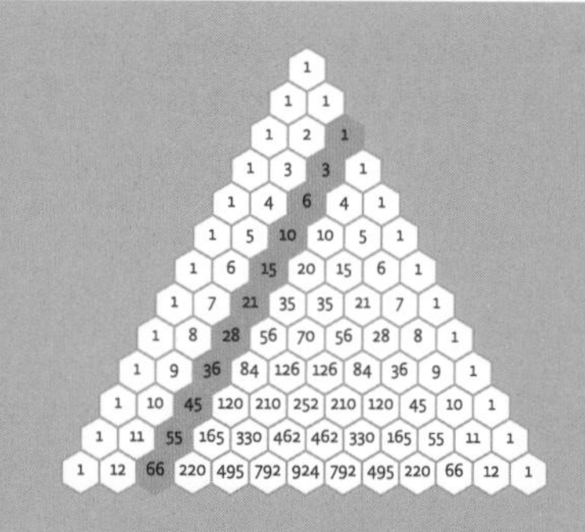

TRIANGULAR NUMBERS

The **next diagonal sequence in from the natural numbers gives the triangular numbers** (formed by summing the dots in a triangle with increasing rows)—1, 3, 6, 10 ...

FIBONACCI

Summing along slightly skewed diagonals produces the **Fibonacci sequence**.

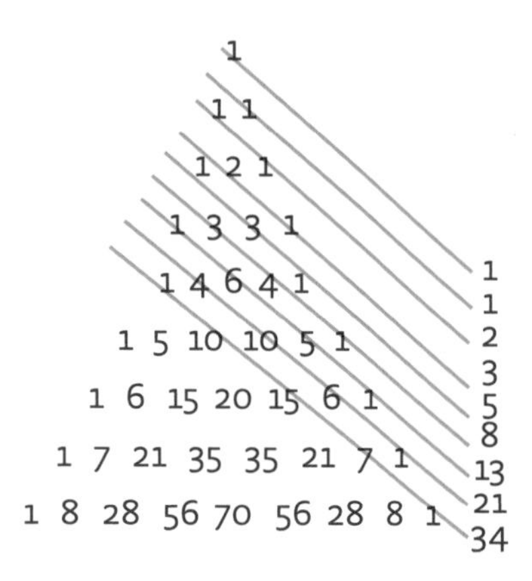

EXPANDING BRACKETS

Expanding the bracket $(1 + x)^2$, gives $1 + 2x + x^2$. Similarly, $(1 + x)^3$ is $1 + 3x + 3x^2 + x^3$. If you call the apex of Pascal's triangle row 0 then, **in general, the expansion of $(1 + x)^n$ will comprise ascending powers of x each multiplied by the numbers from the triangle's nth row**.

$$(1+x)^2 = 1 + 2x + x^2$$

$$(1+x)^3 = 1 + 3x + 3x^2 + x^3$$

$$(1+x)^4 = 1 + 4x + 6x^2 + 4x^3 + x^4$$

$$(1+x)^5 = 1 + 5x + 10x^2 + 10x^3 + 5x^4 + x^5$$

$$(1+x)^6 = 1 + 6x + 15x^2 + 20x^3 + 15x^4 + 6x^5 + x^6$$

COMBINATORICS

If the top of the triangle is row 0, and the outer left edge is column 0, then the number of ways to choose k objects from n is given by row n column k. For example, there are six unique ways to pull two marbles from a bag of four—as given by column 2, row 4.

ANCIENT KNOWLEDGE

Although named after the seventeenth-century French mathematician Blaise Pascal, the triangle was known to Persian mathematicians, including Omar Khayyam, as far back as the eleventh century.

PROBABILITY

A way to get a grip on the outcome of random events, probability can be used by gamblers, physicists, and anyone deciding whether to take an umbrella with them.

$$P(X) = \frac{\text{Number of outcomes where X happens}}{\text{Total number of possible outcomes}}$$

- **1650s** Fermat, Pascal, Cardano, and Huygens lay the foundations of probability theory
- **1718** Abraham de Moivre's *The Doctrine of Chances* is published
- **1761** Bayes's theorem on conditional probability is proved
- **1812** Pierre-Simon Laplace publishes his book *Analytic Theory of Probabilities*
- **1859** James Clerk Maxwell formulates the first ideas in statistical mechanics, applying probability to physics

PLACE YOUR BET

Flip a fair coin and the **probabilities of heads or tails** are both ½, or 0.5.

Roll a six-sided die and **the probability of any particular number coming up** is 1/6, or 0.167.

In general, the probability of some event X happening, denoted P(X), is given by

P(X) = (Number of outcomes where X happens)/(Total number of possible outcomes)

For example, if you rolled two six-sided dice, there are thirty-six possible outcomes—all equally likely.

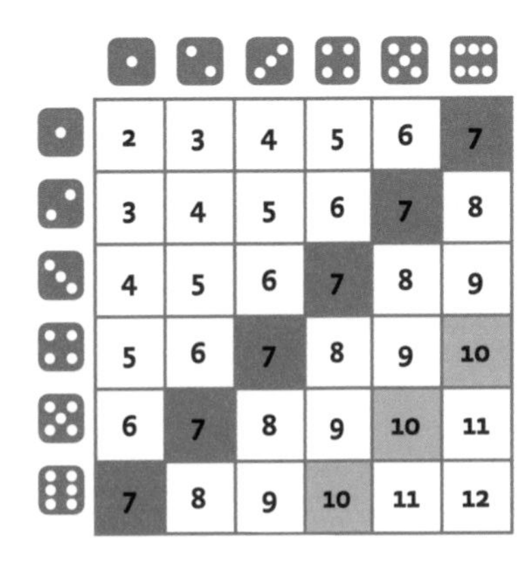

	⚀	⚁	⚂	⚃	⚄	⚅
⚀	2	3	4	5	6	7
⚁	3	4	5	6	7	8
⚂	4	5	6	7	8	9
⚃	5	6	7	8	9	10
⚄	6	7	8	9	10	11
⚅	7	8	9	10	11	12

Of these, six outcomes add up to 7. So your probability of getting 7 is 6/36 = 1/6. On the other hand, your probability of rolling 10 is just 3/36 (since there are only three ways of doing this) = 1/12 .

RULES OF PROBABILITY

If two events are **independent**—that is, one has no bearing on **the other—then the probability of both happening is given by multiplying their individual probabilities together**.

$$P(X \text{ and } Y) = P(X) \times P(Y)$$

On the other hand, if two events are **mutually exclusive**—meaning that one rules out the other—then **the probability of either one or the other happening is given by adding their individual probabilities together**.

$$P(X \text{ or } Y) = P(X) + P(Y)$$

INFINITY

Infinity is more than just a very big number. Mathematicians have spent centuries grappling with its head-bending subtleties and paradoxes.

EARLY MODERN

6th century BC Anaximander puts forward the notion of infinity

5th century BC Eudoxus of Cnidus considers the opposite of infinity—the infinitesimally small

4th century BC The ancient Greeks Plato and Aristotle both abhor the notion of the infinite

1657 John Wallis introduces the "love knot" as the symbol for infinity

1874 Cantor carries out his work on the concept of infinity and cardinality

WHAT IS INFINITY?

Infinity **does not behave as you might expect**. Imagine the largest number you can think of. Now add 1 to it. The new number you have is one bigger than the number you had before. **However, $\infty + 1$ is still just ∞.**

CARDINALITY

The **German mathematician Georg Cantor** showed that there are **different levels of infinity**. He split these up according to what he called "cardinality," labeled by the Hebrew letter ℵ ("aleph").

1,2,3...

The ground level is $\aleph_0$, and **contains all the integers, odd numbers and even numbers**. These are known as the **"countable numbers." Whereas you might think that the infinity of odd and even numbers should be twice the size of the infinity of integers, they are—bizarrely—all the same.**

1/2, 1/3, 1/4...

Cantor showed that the **rational fractions** also reside in $\aleph_0$.

Higher ℵ levels exist to classify higher infinities.

Irrational numbers such as π, e, $\sqrt{2}$ form the **"continuum,"** and have a mind-boggling cardinality of $2^{\aleph_0}$.

CALCULUS

Calculus is a method of analysis, concerned with finding the gradients of and area under mathematical curves. It has applications in everything from quantum physics to economics.

DIFFERENTIATION

Imagine you're driving up a hill. The steeper the slope, or **"gradient"** of the hill, the harder your car's engine has to work to climb it.

Differentiation is a **mathematical technique for calculating the gradient of a curve**. It's equivalent to working out the **tangent**, the straight line touching the curve at any point along its length.

HOW IT WORKS

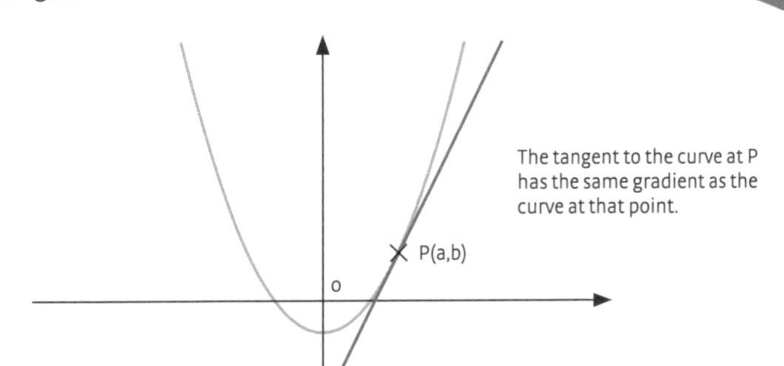

The tangent to the curve at P has the same gradient as the curve at that point.

If you have an equation $y = x^n$, then the gradient is given by nx^{n-1}. The gradient of y with respect to x is usually written as $\frac{dy}{dx}$, a piece of **notation** introduced by **Gottfried Leibniz**. Some common mathematical functions differentiated are ...

$$\frac{d}{dx} n = 0$$

$$\frac{d}{dx} x = 1$$

$$\frac{d}{dx} x^n = nx^{n-1}$$

$$\frac{d}{dx} e^x = e^x$$

$$\frac{d}{dx} \ln x = \frac{1}{x}$$

$$\frac{d}{dx} n^x = n^x \ln x$$

$$\frac{d}{dx} \sin x = \cos x$$

$$\frac{d}{dx} \cos x = \sin x$$

INTEGRATION

The **inverse is integration**.
The **integral** of an equation gives the total area under the curve.
It **works in the exact opposite way to differentiation**.
So, if $y = x^n$ then the integral of y is

$$\int y\, dx = \int x^n dx = \frac{x^{n+1}}{n+1} + C$$

Where dx shows that the integration is being done with respect to x, and C is a **numerical constant** which must be added in general (and which disappears when you differentiate to get back the original equation).

APPLICATIONS

Minima and maxima: differentiation is used to locate the largest and smallest values of a function.

Area and volume: integration can be used to calculate **areas and volumes of shapes and solids**.

Dynamics: calculus is applied extensively in the **physics of time-varying phenomena**.

Numerical analysis: estimates of the **slope of a function** can be used to find **numerical solutions**, usually with the help of a **computer**.

GOTTFRIED WILHELM LEIBNIZ

An exceptionally gifted polymath, Leibniz was the first scholar to publish a theory of calculus. He also made contributions to geometry, topology, and solving equations.

July 1, 1646 Leibniz is born in Leipzig, Saxony, where his father is a professor of moral philosophy. A child prodigy, he enjoys delving into his father's library of philosophical works

1661 At the age of fourteen, he enrolls at Leipzig University to study philosophy. A year later he earns his bachelor's degree, before completing his master's in 1664. He also studies law

1666 Leibniz begins a long career as a political assistant and diplomat, working across Europe. While living in Paris, he teaches himself mathematics and physics

1673 Leibniz has been working on the prototype of a calculating machine that could add, subtract, multiply, and divide. While in London, he demonstrates it to the Royal Society, which quickly makes him a member

1675 Leibniz develops his own version of integral and differential calculus

1684 Leibniz publishes his findings on calculus in the book *Nova Methodus pro Maximis et Minimis*. In later works, he recognized the contribution that Newton had made in the field

1712 An outcry erupts over who invented calculus. A group of scholars accuses Leibniz of plagiarizing Newton's work. It's believed that Newton was behind this

November 14, 1716 Leibniz never recovers from the slander and dies in Hanover at age seventy. In posterity, both Leibniz and Newton are credited with inventing calculus. However, Leibniz's notation was favored over Newton's and is still in use today

ISAAC NEWTON

One of the most brilliant scientists of all time, Newton advanced every area of mathematics he studied. His work on calculus is the bedrock of physics and engineering.

BRITAIN'S GREATEST SCIENTIST

According to the Julian calendar, Newton was **born on Christmas Day, 1642**, at **Woolsthorpe Manor**, Lincolnshire. His family members were farmers

Newton's **father** died three months before he was born and his **mother** remarried, leaving him to be raised by his **grandparents**

In 1661, Newton went up to **Trinity College**, Cambridge, to study **law**. He **worked as a valet** to fund his studies

Plague rampaged through Cambridge in 1665, so Newton returned to Woolsthorpe where he spent two years pondering **calculus**, **optics**, **planetary motion**, and **gravity**

Newton was made a **fellow of Trinity College, Cambridge**, and in 1668 he succeeded **Isaac Barrow** as the prestigious **Lucasian Professor of Mathematics**

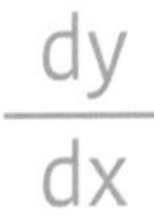

Newton developed the **theory of calculus** at around the same time as German mathematician **Gottfried Leibniz**, leading to a **dispute over who had actually discovered it**

His genius was recognized in his lifetime and he was appointed to a number of prestigious positions, including **Master of the Royal Mint**

Newton **died** in his sleep on March 31, 1727, in Kensington, London. His body was buried at **Westminster Abbey**

FACTS AND FIGURES

He was knighted by Queen Anne in **1705**, the second scientist to receive the honor.

Newton served **two** years as Member of Parliament for Cambridge University.

He spent **twenty-four** years as president of the Royal Society.

In **1687**, his seminal work *Philosophiae Naturalis Principia Mathematica* was published.

NUMERICAL ANALYSIS

While some mathematical equations are easy to solve, others are fiendishly complex, with solutions that can only be found by the brute-force techniques of numerical analysis.

1656 John Wallis publishes numerical methods to approximate the area under curves—essentially numerical integration

1675 Isaac Newton first publishes his method for solving equations numerically

1690 Joseph Raphson refines and simplifies Newton's method

1895 Carl Runge publishes a numerical method for solving differential equations—those involving both a variable and its gradient

1940s Invention of the electronic computer means numerical analysis can be automated

EARLY MODERN

GRAPHICAL SOLUTIONS

The equation $x - 2 = 0$ is easily solved by adding 2 to both sides, giving the solution $x = 2$. But what about $x^5 + \sqrt{x} - 3 = 0$? Equations like this have no simple pen-and-paper solution. Instead, mathematicians apply **algorithms** to obtain solutions numerically.

One simple approach is to **plot the function graphically**.

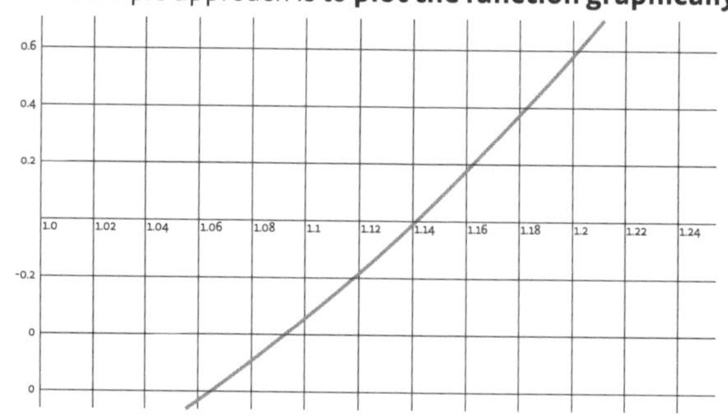

The graph shows that $x^5 + \sqrt{x} - 3$ passes through zero when $x \approx 1.14077$.

INTERVAL BISECTION

If you know that the solution to an equation lies in a given interval, then **interval bisection** is a powerful numerical technique that **works by dividing the interval in half, finding which half contains the solution**, and then **repeating**.

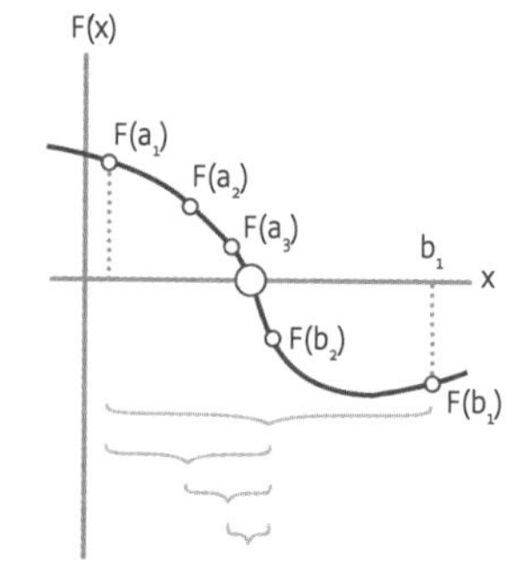

NEWTON'S METHOD

Isaac Newton (and, independently, the English mathematician **Joseph Raphson**) developed a **numerical technique for solving equations** in the late seventeenth century.

To solve $f(x) = 0$, Newton's method involves **starting with a guess** (x_0), calculating the **tangent to the curve** at that point, finding where the **tangent line is zero** (x_1), and then using this as the **next guess**. The process is **repeated until the estimates converge** on the solution.

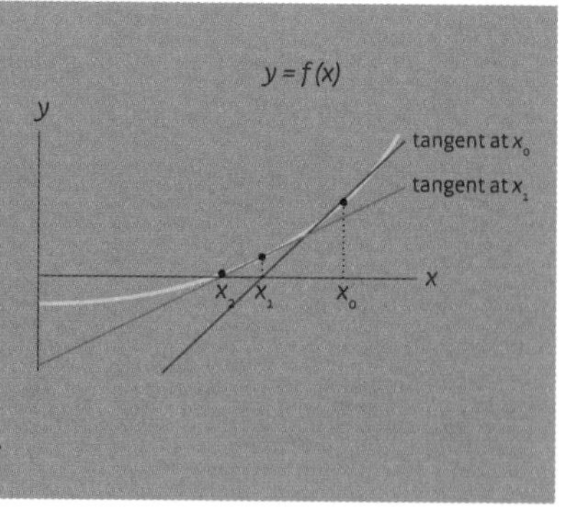

MECHANICS

One of the earliest applications of mathematics to the physical world, mechanics deals with the motion of bodies under the application of forces.

- **4th century BC** The ancient Greeks consider how objects react to forces
- **1687** Newton publishes his three laws of motion
- **1765** Leonhard Euler extends the laws of classical mechanics to rotating bodies
- **1788** Lagrange develops an alternative approach to mechanics, based on energy rather than forces
- **1900** Max Planck develops the first ideas in quantum mechanics, for the motion of subatomic particles
- **1905** Einstein's special theory of relativity is published, showing how Newton's laws break down at close to the speed of light
- **1928** Paul Dirac combines quantum mechanics with special relativity to describe ultrafast motion in the subatomic world

WHAT IS IT?

Mechanics describes **billiard balls colliding**, **swinging pendulums**, **blocks sliding down slopes**, **systems of pulleys**, and much more.

Mechanics was governed for hundreds of years by **Newton's laws of motion**, but in the twentieth century **corrections** were introduced for **objects that are traveling very fast (relativity theory)** or that are **very small (quantum mechanics)**.

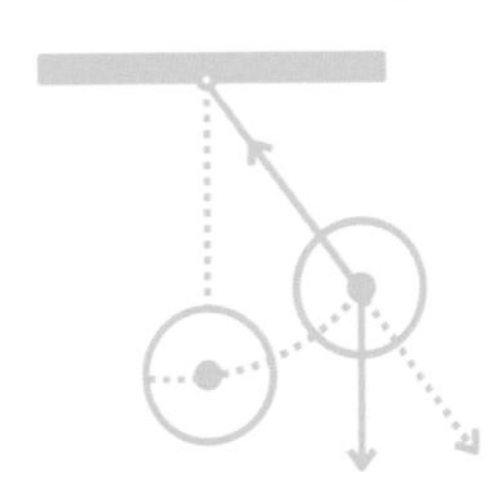

NEWTON'S LAWS

British scientist **Sir Isaac Newton** encapsulated the principles of **classical mechanics** in his **three laws of motion:**

1. A body remains at rest or continues in its state of uniform motion unless acted on by a force.

2. If you apply a force F to a body of mass m, it will accelerate at a rate a, satisfying the equation $F = ma$.

The more force, the more acceleration

3. For every action there is an equal and opposite reaction. That's why a rifle kicks back against your shoulder when you fire it.

POLAR COORDINATES

Rather than a grid reference, polar coordinates give your position as a distance and a direction from a fixed point. They are used every day in navigation, robotics, mathematics, and physics.

WHAT ARE POLAR COORDINATES?

You may have used a form of polar coordinate yourself. If someone asks where you live, you might say "twenty miles north of London." What you are doing is giving a **rough set of polar coordinates**Ò a **direction** and a **distance**.

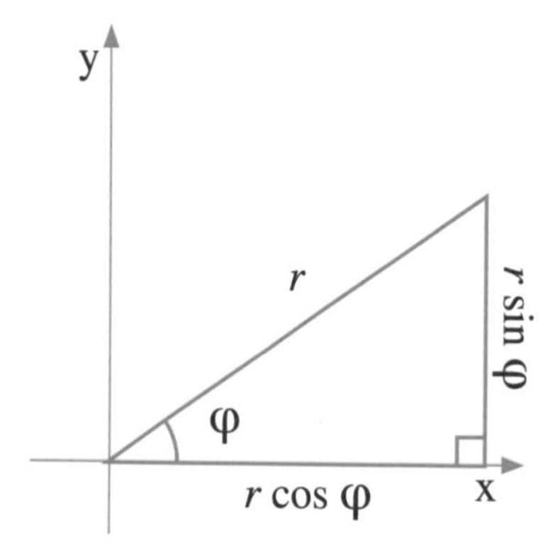

The diagram shows a 2-D polar coordinate system, in which a point on a 2-D plane is specified by the **angle** φ, measured counterclockwise from the *x* **axis**, and the **radial distance** *r*. Polar coordinates can be related to **Cartesian** (x, y) **grid coordinates** by the equations.

$$x = r \cos \varphi$$

$$y = r \sin \varphi$$

ARCHIMEDEAN SPIRAL

Shapes can be defined by polar coordinates, including **spirals**, and **cycloids**. **Archimedes** described a spiral whereby the **radius** increases in proportion to the **angle**.

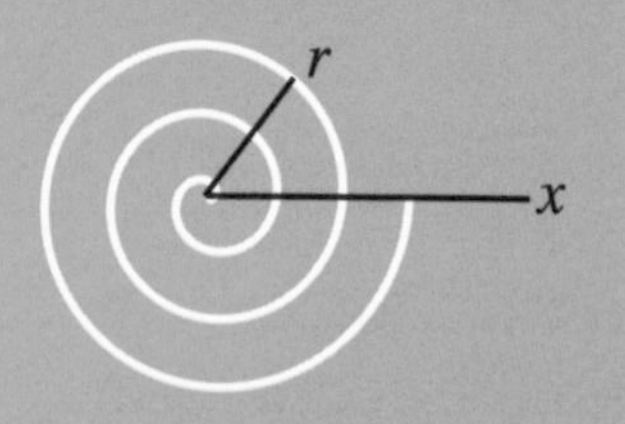

The diagram shows an Archimedean spiral where *r* increases linearly from zero as φ sweeps counterclockwise from the *x* axis.

DID YOU KNOW?

Concepts of angle and radius were used by ancient thinkers. **Hipparchus** (c. second century BC) used polar coordinates to establish the position of celestial bodies.

Polar coordinates greatly simplify the **equations of motion** of **rotating mechanical systems**.

3-D polar coordinates are used in the **physics of electric, magnetic, and temperature fields**.

In 1691, the Swiss mathematician **Jacob Bernoulli** developed polar coordinates for a wide range of mathematical problems.

They can be used to measure **planetary motion**.

ABRAHAM DE MOIVRE

Abraham de Moivre is renowned for his eponymous formula, which links complex numbers and trigonometry, and for advancing the study of probability.

LIFE AND TIMES

Born in the Champagne province of France on May 26, 1667. De Moivre studied **physics** in Paris but due to his faith (he was a **French Protestant**) he was forced to **flee persecution** and settle in **London**. Here, he became friends with **Sir Isaac Newton** and the astronomer **Edmond Halley**.

$$(\cos x + i \sin x)^n = \cos(nx) + i \sin(nx)$$

DE MOIVRE'S FORMULA

For any **real number** x and **integer** n ...
$(\cos x + i \sin x)^n = \cos(nx) + i \sin(nx)$.

The theorem gives a formula for calculating the **powers of complex numbers**. These are numbers that can be written in the form a + bi, where i is the **unit imaginary number** (defined such that $i^2 = -1$).

De Moivre's formula connects **complex numbers** and **trigonometry**. He was instrumental in bringing trigonometry out of the realm of **geometry** and into **mathematical analysis**.

PROBABILITY

Despite his brilliance, de Moivre couldn't find a chair in mathematics at a university. He earned a living as a **private mathematics tutor** and a **consultant on gambling and insurance**. His 1718 work ***The Doctrine of Chances*** covered a wide range of problems in dice and other games. It greatly enhanced the study of probability theory.

DID YOU KNOW?

De Moivre is reputed to have **predicted the date of his own death**—successfully! He noticed that in old age he was sleeping an extra fifteen minutes each night and correctly calculated that his sleep time would be twenty-four hours a day on November 27, 1754, the date of his actual death.

INFINITE SERIES

The ability to write mathematical functions as series—an often-infinite sum of terms—is a powerful trick that can be used in a host of applications.

SUMMING THE INFINITE

You might think an **infinite sum** of terms would be **infinite in value**. But imagine a runner approaching the finish line of a race. You could break their motion down into an infinite series of steps, each traveling half the remaining distance to the finish. Despite the fact that there are an **infinite number of steps**, the runner still crosses the line in a **finite time**.

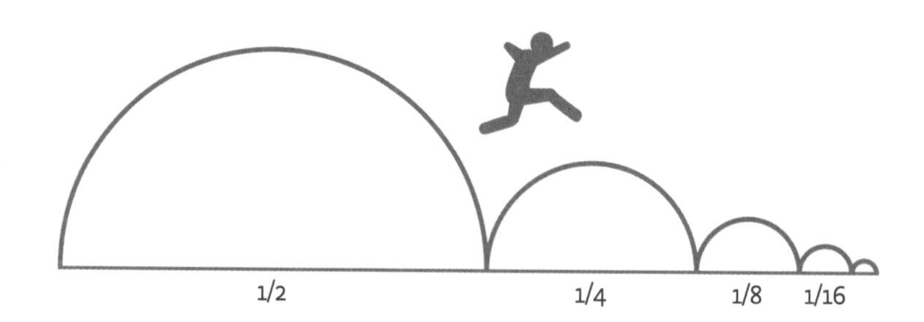

The fraction of the distance to the finish that they cover with each step obeys the infinite series

$$\frac{1}{2} + \frac{1}{4} + \frac{1}{8} + \frac{1}{16} + \ldots$$

which converges to the value 1.

NOTATION

Sums are denoted by the capital **Greek letter** Σ (pronounced "sigma"), with a **subscript** giving the **lower limit** of the series and a **superscript** giving the **upper limit**. For example,

$$\sum_{k=0}^{\infty} x^k$$

means the sum from $k = 0$ to infinity of x^k.

EXAMPLES

Some series with well-defined sums

$$\sum_{k=1}^{\infty} \frac{1}{k^2} = \frac{\pi^2}{6}$$

$$\sum_{k=1}^{\infty} \frac{x^k}{x!} = e^x$$

TAYLOR SERIES

Taylor series are named after the English mathematician **Brook Taylor**, who showed in 1715 that any mathematical function $f(x)$ can be approximated in the neighborhood of a point $x = a$, by the series

$$f(x) \approx f(a) + \frac{1}{1!}\frac{df(a)}{dx}(x-a) + \frac{1}{2!}\frac{d^2f(a)}{dx^2}(x-a)^2 + \ldots$$

Where ! is the factorial function, $\frac{df(a)}{dx}$ is $f(x)$ differentiated with respect to x and evaluated at $x = a$, and $\frac{d^2f(a)}{dx^2}$ is $f(x)$ differentiated twice with respect to x and evaluated at $x = a$.

For example, $f(x) = x^2$ can be approximated by

$$f(x) \approx a^2 + 2a(x-a) + (x-a)^2 + \ldots$$

The approximation holds for values of x close to a, so the higher powers of $(x - a)$ soon become small enough to be ignored.

GRAPH THEORY

Networks of points and the connections between them can be analyzed using a branch of mathematics known as graph theory.

THE BRIDGES OF KÖNIGSBERG

In the eighteenth century, the city of **Königsberg**, built straddling the River Pregel in modern-day Russia, was composed of four disjointed bodies of land, connected by seven bridges.

The question was posed: is it possible to find a path around Konigsberg so that you cross each bridge once and only once?

ABSTRACT GRAPHS

In 1735, Swiss mathematician **Leonhard Euler** had the brainwave of **reducing the map of Königsberg to an abstract form** where the **landmasses** were represented as simple **dots** and the **bridges** were shown as lines connecting them. The lengths of the **lines** were unimportant—all that mattered was the **structure of the connections**.

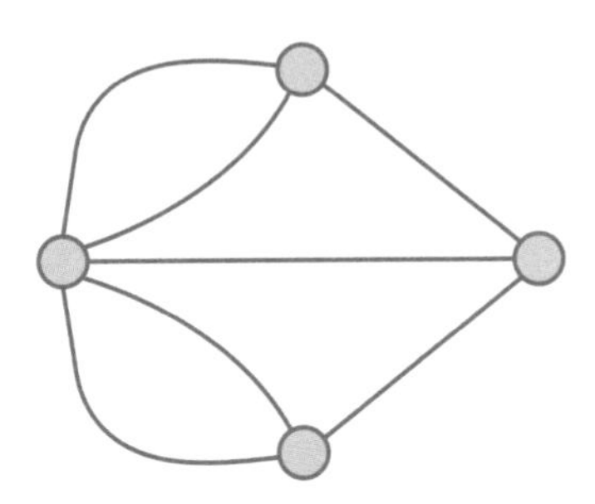

The question then becomes: starting at one of the four points (or "**nodes**," as they're called) can you find a route around this "**graph**" that traverses each connection (or "**edge**") just the once?

EULER'S SOLUTION

Euler realized that a route can only be found if, after discounting the start and end, the remaining nodes all have an **even number of edges connecting to them**—that is, they are of **"even degree."** This has become known as **Euler's theorem**. In the case of Königsberg's bridges, all the nodes have **odd degree** and so the feat is impossible. Try it. This was the **first application** of what has become known as **"graph theory."**

Some very complicated graphs are possible.

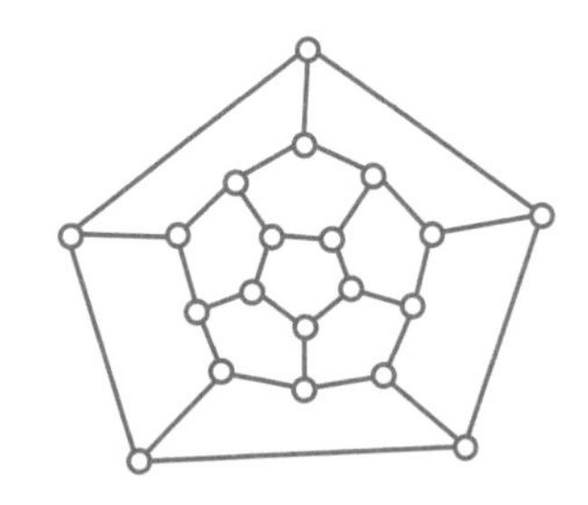

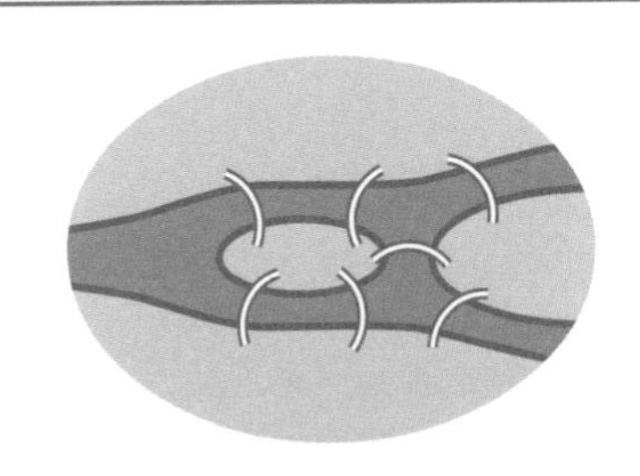

OTHER APPLICATIONS

Graph theory is also used in:

- **Topology**
- **Statistical modeling**
- **Computing**
- **Group theory and sciences**

GAMBLING

In the eighteenth century, there were handsome profits to be made from applying math to games of chance. Some mathematicians supplemented their incomes this way.

EXPECTED VALUE

A gambler makes money by maximizing the **"expected value"** of their bets. This is given by **summing up profit** and loss in every possible outcome **multiplied by the probability of each**.

EXAMPLE

Imagine a casino game with a fair six-sided die. The stake is \$6, which you lose if the roll is one to five, but if a six comes up then you win \$24. In this case, the expected value is

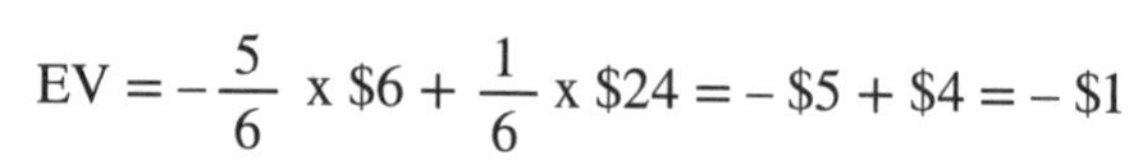

$$EV = -\frac{5}{6} \text{ x } \$6 + \frac{1}{6} \text{ x } \$24 = -\$5 + \$4 = -\$1$$

So, every time you play, on average, you lose \$1. Avoid this.

If, on the other hand, the cost to play dropped to \$3, then the EV rises to +\$1.50 and you're onto a winner!

ST. PETERSBURG PARADOX

Now imagine a coin-tossing game—tails you lose \$2, heads you win \$2. Ordinarily, this game has an EV = 0. But what if, every time you lose, you double your stake, only resetting once you finally win? Your EV for each "play" is...

$$EV = -\frac{1}{2} \text{ x } \$2 + \frac{1}{2} \text{ x } \frac{1}{2} \text{ x } \$4 + \frac{1}{2} \text{ x} \frac{1}{2} \text{ x} \frac{1}{2} \text{ x } \$8 + \ldots = \$1 + \$1 + \$1 + \ldots$$

Which is **infinite**.
This is known as the **St. Petersburg Paradox**, after the Russian city where mathematician **Daniel Bernoulli** was living when he attempted to solve it.

THE SOLUTION

Bernoulli believed the solution lay in the notion of **utility**—the **diminishing value of money**. If you're given \$1 million, for example, it may change your life. But would \$2 million change your life twice as much? Perhaps not. That's utility.

The modern resolution lies in the fact that **you'd need an infinite bankroll to play this strategy**, to cover the possible inordinately long runs of tails—and that the casino would eventually stop accepting your **spiraling stakes**.

LEONHARD EULER

Euler's work advanced the fields of geometry, algebra, number theory, trigonometry, and infinitesimal calculus. Mathematics is replete with formulas and theorems named after him.

WHO WAS LEONHARD EULER?

Euler (pronounced "oiler") was born in Basel, Switzerland, on April 15, 1707. His **father** was a **pastor** of the Reformed Church. The family was close to **Johann Bernoulli**, one of the greatest mathematicians in Europe.

In 1720, Euler attended the **University of Basel** and later studied for a master's on the works of **Descartes** and **Newton**.

Euler was a **professor of physics** at the **Imperial Russian Academy of Sciences** in **St. Petersburg**. In 1741, he joined the **Berlin Academy**.

Euler's **eyesight deteriorated** through his career. After suffering a fever in 1738, he became almost **blind in his right eye**. This had little effect on his **productivity**, thanks to his **remarkable calculation skills** and **astonishing memory**.

Euler **died of a brain hemorrhage** on September 18, 1783 in St Petersburg, Russia.

THE MOST BEAUTIFUL EQUATION

Euler's identity is hailed as the **most beautiful equation in mathematics**. Some people even have it **tattooed on their bodies**. Why?

$$e^{i\pi} + 1 = 0$$

The equation is said to hold **mathematical beauty** because it includes **three key functions: addition**, **multiplication**, and **exponentiation**. It also links **five mathematical constants:** 0, 1, π, e (the base of natural **logarithms**, 2.7182818284), and i (the **unit imaginary number**).

As such, it presents a **deep connection** between the realms of **geometry**, **calculus**, and **complex numbers**.

Some historians have suggested that Euler himself wasn't the first to derive the identity, though it certainly **emerges from concepts he helped to develop**.

TOPOLOGY

Topology is the study of surfaces and solids when they are twisted, stretched, and deformed—but not cut or torn. It has applications in studies of knots, particle physics, and the universe at large.

EARLY MODERN

300 BC Euclid publishes *Elements*, containing the basic postulates of geometry

1736 Leonhard Euler publishes the first work on graph theory, an early forerunner to topology

1858 Möbius and Listing discover the "Möbius strip"

1882 Felix Klein discovers the Klein bottle

2006 Russian mathematician Grigori Perelman proves the Poincaré conjecture, concerning the topology of a 3-D sphere

FLEXIBLE GEOMETRY

As far as a topologist is concerned, **two shapes are the same if one can be deformed into the other**. Imagine they are made of perfectly stretchy rubber. That means a **sphere is topologically the same as a cube**, because one can be pulled or squashed into the shape of the other, and vice versa.

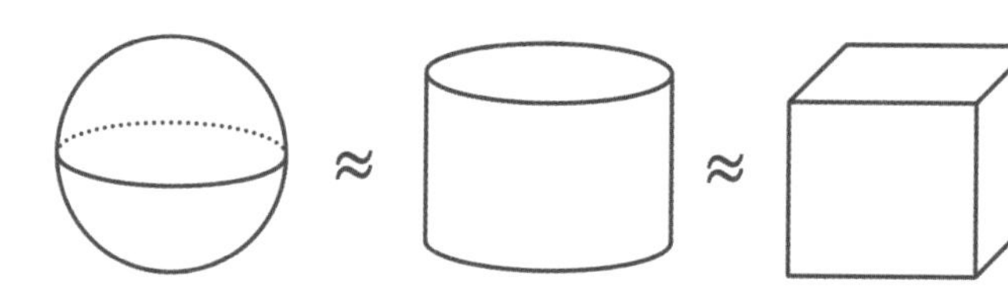

THE HOLE TRUTH

Not being allowed to tear or cut means that **topologically equivalent objects must have the same number of "holes."** Cut a hole in a sheet of paper and there's no way it can be transformed into a sheet with no hole. **Objects with the same number of holes can be morphed into one another though.** So, to a topologist, a donut and a tea cup are one and the same!

WEIRD FIGURES

The **Möbius strip** (below) is something of a **topological oddity**—it is a **surface with only one side**.

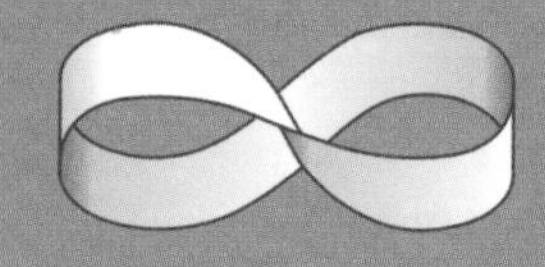

It's made by taking a **rectangular strip**, adding a **180-degree twist**, and then **gluing the ends together**.

The Möbius strip still has a boundary along its edge; however, the **Klein bottle** (right) is an example of a **one-sided surface with no boundary**.

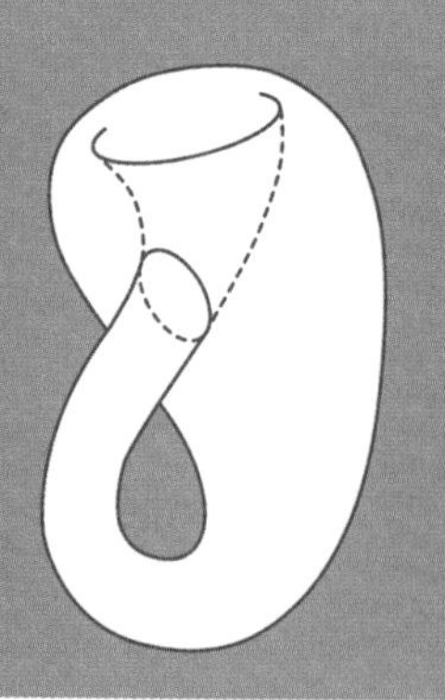

THE BERNOULLI FAMILY

This Swiss dynasty produced no fewer than eight gifted mathematicians, who between them made enormous contributions to the study of both mathematics and theoretical physics.

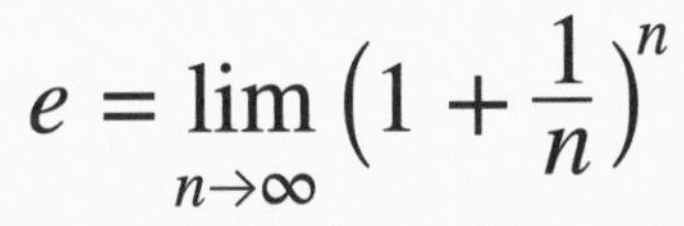

$$e = \lim_{n\to\infty}\left(1 + \frac{1}{n}\right)^n$$

Jacob Bernoulli (1654–1705) wrote the work ***Ars Conjectandi*** on probability. **"Bernoulli numbers"** are named in his honor and he discovered the **mathematical constant *e*.** It's found in many mathematical formulas.

Johann Bernoulli (1667–1748) Brothers Jacob and Johann worked together on **infinitesimal calculus** and its applications, including the **mathematics of ship sails** and **optics**.

Nicolaus Bernoulli (1662–1718) The third Bernoulli brother was a painter and alderman (essentially a council member) in the Swiss city of Basel.

Nicolaus I Bernoulli (1687–1759), nephew to Jacob and Johann, specialized in **probability theory**, **geometry**, and **differential equations**.

Daniel Bernoulli (1700–82), second son of Johann, applied mathematics to **fluid mechanics** (discovering the **principle by which aircraft would later fly**) and carried out **ground-breaking research** into **statistics** and **probability theory**.

Johann II Bernoulli (1710–90), third son of Johann, succeeded his father as **professor of mathematics** at the **University of Basel** and also **studied physics**.

Nicolaus II Bernoulli (1695–1726), son of Johann, invented the **St. Petersburg Paradox** related to **probability in economics** and **gambling**.

Johann III Bernoulli (1744–1807), son of Johann II, was a **child prodigy**. By the age of nineteen, he was **Astronomer Royal of Berlin**. He taught mathematics at **Berlin Academy** and wrote works on **astronomy**, **travel**, and **geography**.

Jacob II Bernoulli (1759–89), third son of Johann II, was born in Russia and became a **professor of mathematics** at the **Academy of St. Petersburg**. He **drowned** while swimming in the River Neva, just a few months after his **marriage to a granddaughter of Leonhard Euler**.

PROBABILITY DISTRIBUTIONS

A probability distribution encapsulates the probabilities of all the possible outcomes of a particular random event—such as a dice roll, a horse race, or tomorrow's weather.

1755 Thomas Simpson studies distributions as a way to understand the errors in experimental measurements

1774 Pierre-Simon Laplace suggests probability distributions based on powers of the exponential function, *e*

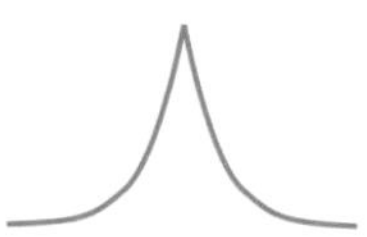

1809 Carl Friedrich Gauss publishes the most important distribution of all, the normal distribution

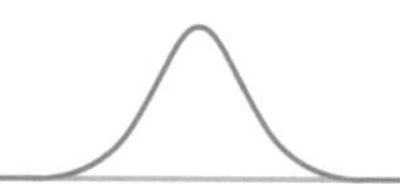

1894 English statistician Karl Pearson introduces the term "standard deviation" to describe the spread of possible values in a probability distribution

SNAKE EYES

Roll two six-sided dice, and the outcome will be a number between 2 and 12. But the **outcomes are not all equally likely**. From the thirty-six possibilities (6 x 6), a score of two can occur just one way (two ones), giving a probability of 1/36, while seven can occur six ways (a one and six, two and five, three and four, four and three, five and two, six and one), so the probability is 6/36, or 1/6.

The probability of every possible score is:

This is an example of a **probability distribution**.

DISCRETE DISTRIBUTIONS

The distribution of dice rolls is a **"discrete distribution,"** because only **whole-number "integer"** outcomes are allowed. Discrete distributions resemble **bar charts**. On the other hand, a **"continuous distribution" permits non-integer values**. In this case, it **makes no sense to talk about the probability of a particular value occurring**. Instead, statisticians work out the probability of a **range of values**, given by **integrating** (that is, applying **integral calculus** to) the **distribution**. For instance, the shaded area in the continuous distribution to the right gives the probability of the random variable *x* as lying between 11 and 14.

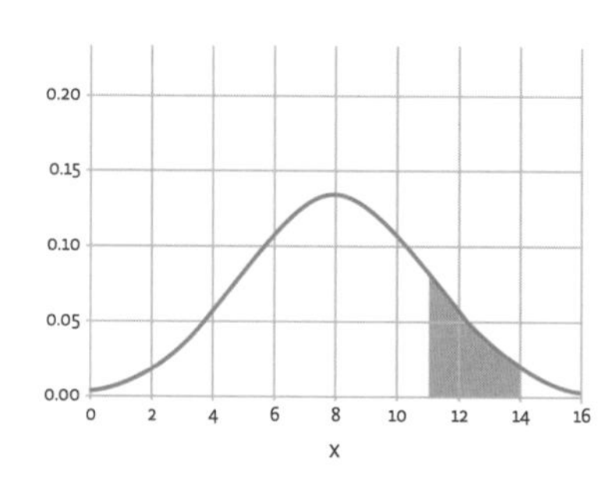

Some commonly occurring probability distributions are shown below.

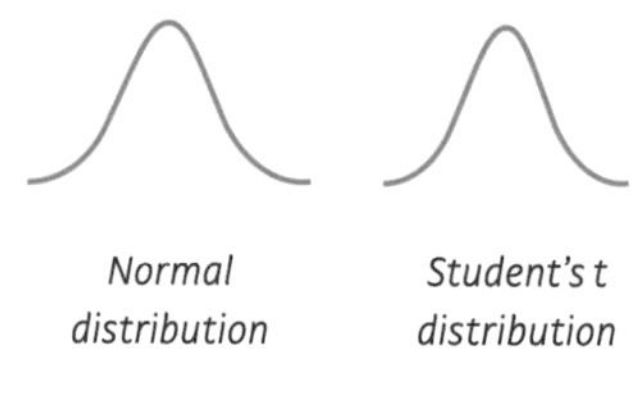

Normal distribution *Student's t distribution*

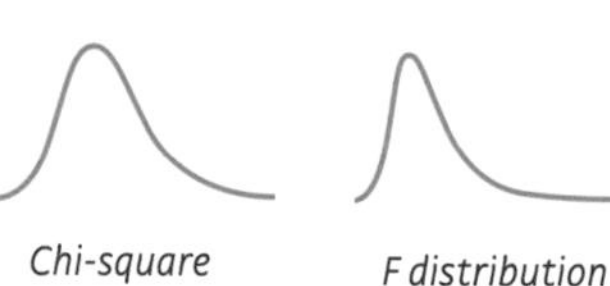

Chi-square distribution *F distribution*

BAYES'S THEOREM

Bayes's theorem, first stated in the eighteenth century by Thomas Bayes, is a mathematical principle about the probabilities of related events occurring.

THOMAS BAYES

1701 Born in London, England

1742 Elected a Fellow of the Royal Society

1761 Dies in Tunbridge Wells. Buried in Bunhill Fields, London

1763 Bayes's *An Essay Toward Solving a Problem in the Doctrine of Chances*, containing the first statement of Bayes's theorem, is read posthumously to the Royal Society

1950s Bayesian statistics finally begins to grow in popularity, having been dismissed in the past as too subjective

WHAT IS IT?

Given two events, labeled *A* and *B*, Bayes's theorem states that **the probability of *A* happening given *B* has happened** is the **same as the probability of *B* given *A* has happened, multiplied by the overall probability of *A* divided by the overall probability of *B*.**

That is,

$$P(A|B) = \frac{P(B|A)\,P(A)}{P(B)}$$

where *P(A|B)* means the probability of *A* given *B*, sometimes called the ***conditional* probability**.

EXAMPLE

A disease affects 2 percent of the population. If you have the disease, a test will detect it 90 percent of the time. The test throws up a positive result 3 percent of the time. That is,

$$P(\text{disease}) = 0.02$$
$$P(\text{positive test} \mid \text{disease}) = 0.9$$
$$P(\text{positive test}) = 0.03$$

Then Bayes's theorem says that the **probability you have the disease given a positive test** is

$$P(\text{disease} \mid \text{positive test}) = \frac{0.9 \times 0.02}{0.03} = 0.6$$

Or 60 percent.

APPLICATIONS

Evidence: used in **court** for assessing the **probability of guilt** given the **available evidence**.

Science: a **Bayesian approach** can be used to **weigh up the results of scientific experiments against prior beliefs**.

Spam filters: these assess the **likelihood of an e-mail being spam** given its content, based on a **Bayesian algorithm**.

KNOT THEORY

A kind of topology that deals with the classification of tangled line segments, knot theory has applications in molecular biology and computer science.

WHAT IS IT?

Knot theory is the **study of mathematical knots**—which are a bit like knots in string, except the **ends of the knot are joined**. As in topology, **two knots are considered equivalent if they can be transformed into one another by twisting**, **stretching**, and **deforming**—but **not cutting**.

THE TREFOIL KNOT

The simplest **non-trivial knot** has **Alexander-Briggs designation** 3_1, and is also known as the **"trefoil" knot**. You can make one by tying a simple **"granny knot"** in a length of string and then **gluing the ends together**.

ALEXANDER-BRIGGS NUMBERS

Mathematical knots are **classified by their number of crossing points**.

Knots in the diagram below are labeled using **Alexander-Briggs notation** – the large number gives the number of crossing points, and the **subscript** is just a label used to distinguish between knots in each category.

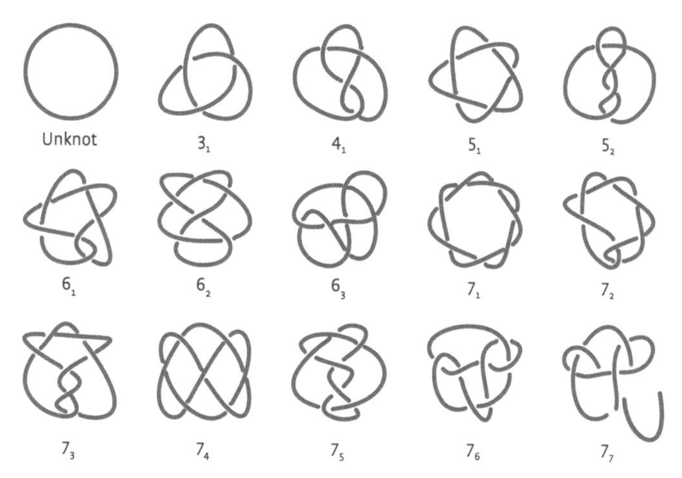

APPLICATIONS

Topological quantum computers: these devices use the braiding together (i.e., **knotting**) of the **tracks left by quantum particles** called **"anyons"** to carry out **computations**.

DNA: studies using knot theory have shown **how cells untangle strands of DNA**.

Shoelace tying: no joke, knot theory can help you tie your shoelaces quickly and more efficiently!

Molecular chirality: knot theory can determine the **"handedness,"** or **mirror symmetry**, of **molecules**.

LATIN SQUARES

Sudoku puzzles are based on ancient mathematical forms called Latin squares. These have modern-day applications in error-correcting codes used in data transmission.

ANCIENT PUZZLES

A Latin square is an **arrangement of integers** such that **each one occurs *n* times in a square *n* × *n*** and **no number appears twice in any row or a column**. Latin squares are believed to have originated in the **Islamic world**, during **medieval times**.

1	2	3
2	3	1
3	1	2

A Latin square where *n* is 3.

GRECO-LATIN SQUARES

In 1700, the Korean mathematician **Choi Seok-Jeong** was the first person to publish Latin squares in grids of nine. Decades later, **Leonhard Euler** attempted to match two together to make a Greco-Latin square. The challenge with Greco-Latin squares is to **combine figures from both squares into one, where no pair of figures is repeated**.

Aα	Bγ	Cβ
Bβ	Cα	Aγ
Cγ	Aβ	Bα

ERROR CORRECTION

Sudoku grids (9 x 9) **can only have a unique solution if at least seventeen clues are provided**. This illustrates the usefulness of Latin squares. If a grid of eighty-one cells can be solved with only seventeen clues present, this means that those clues provide the **answer to missing or incorrect data**. This applies to modern **data transfer**, where Latin squares and similar objects are used as **error-correcting devices**.

VAST POSSIBILITIES

Latin squares of any size can be constructed and the **number of unique combinations grows rapidly**. For example, there are only two different 2 × 2 Latin squares (that's after allowing for **symmetry—any two squares that can be flipped or rotated onto one another are considered identical)**. Meanwhile, there are twelve different 3 × 3 Latin squares and when it comes to nine—the basis of **Sudoku puzzles**—the **number of possible grids** runs to an enormous 6,670,903,752,021,072,936,960!

JOSEPH LOUIS LAGRANGE

An astronomer, mathematician, and theoretical physicist, Lagrange made contributions to classical mechanics, the physics of orbits, and the theory of partial differential equations.

1736 Lagrange is born on January 25 in Turin, to a family of French origin. His **father** was treasurer to the King of Sardinia, but lost his fortune through speculation. Lagrange remarked that **if he had been rich, he probably would not have devoted his life to mathematics**

1750s He attends the **University of Turin** and studies classical Latin. His interest in mathematics begins with the memoir of British astronomer **Edmond Halley**. He is mostly **self-taught** in the field

1755 While still a teenager, Lagrange begins **teaching math at the military academy of Turin**. His work is well received, in particular by **Leonhard Euler**

1764 He presents a **prize-winning essay** to the **French Academy of Sciences** on the **"libration" of the Moon**, the apparent oscillation of the Moon that causes changes in its features visible from Earth

1766 He accepts a post in **Berlin** at the invitation of **Frederick the Great**, who wants the **"greatest mathematician in Europe at his court"**

1766–87 His time in Berlin proves highly productive, with advances in **prime number theory**, **algebra**, **differential equations**, **probability**, and **mechanics**

1786 After Frederick's death, he is **invited to France** by **Louis XVI**, where he is received with honor and given apartments in the Louvre

1789 **Revolution** breaks out in France, and **non-nationals are ordered to leave**. Lagrange, however, is allowed to stay on and **continue his studies**

1794 He is appointed **professor** of the prestigious **École Polytechnique**

1813 **Lagrange dies** in Paris on April 10

ADRIEN-MARIE LEGENDRE

This ground-breaking French scholar was among those who applied mathematics to physics, with intriguing results.

WHO WAS LEGENDRE?

Born into a wealthy Parisian family in 1752, Legendre studied **mathematics** and **physics** at the **Collège Mazarin**

He **taught mathematics** at the École Militaire from 1775 to 1780 and at the **French Academy of Sciences**, which closed during the French Revolution

In 1782, Legendre won a **prize from the Berlin Academy** to describe the **curve made by cannonballs**, taking into consideration **air resistance**, **velocity**, and **angle of projection**

Under **Napoleon's** reign, he taught at the prestigious **École Polytechnique** and was made an officer of the **Légion d'Honneur**

Legendre **died in Paris in 1833** after a **long, painful illness**. His name is among those of the seventy-two French scientists who are **commemorated on the Eiffel Tower**

ACHIEVEMENTS

In his 1794 work ***Éléments de Géométrie,*** Legendre simplified and reorganized **Euclid's *Elements.*** This became a **leading mathematical textbook for 100 years**.

Legendre carried out extensive research into **elliptic integrals**. These are important in calculating the **arc-length of an ellipse** and have applications in **mechanics**.

Legendre's major work ***Exercices de Calcul Intégral*** (1811) explained the properties of **elliptic integrals** and **beta and gamma functions**.

He is also renowned for the **Legendre transformation**, which is used in **classical mechanics and thermodynamics**, and for solving **differential equations**.

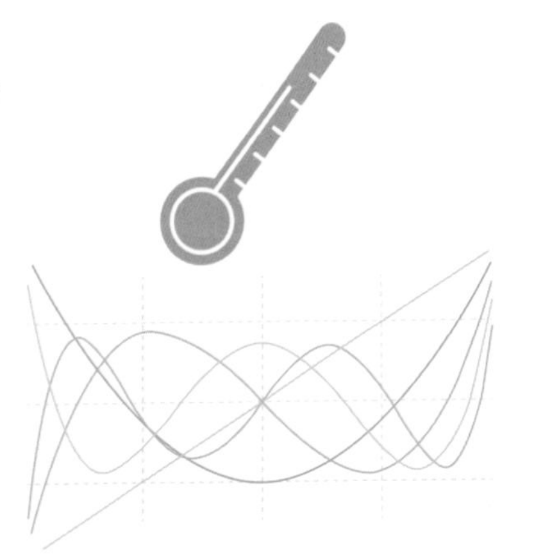

Legendre polynomials are used in **physics** and **engineering**. A polynomial is an **expression of more than two algebraic terms**. These can be expressed in striking forms.

Legendre studied **number theory**, conjecturing the **quadratic reciprocity law**, determining the existence of **integer solutions to quadratic equations**, which was later proved by the German mathematician **Carl Friedrich Gauss**.

PIERRE-SIMON LAPLACE

One of the greatest scientists of all time, Laplace worked on calculus, celestial mechanics, and the theory of tides. He also developed probability theory.

He is born **March 23, 1749**, in Beaumont-en-Auge, Normandy, the son of a cider merchant

In **1765**, Laplace attends Caen University, where his mathematical brilliance shines

Laplace is **nineteen** when he becomes a professor at the École Militaire, where he is Napoleon's examiner

In **1796**, he publishes *Exposition du Système du Monde*, his *magnum opus* on celestial mechanics

Napoleon makes him a count in **1806** and he becomes a marquis in **1817** after the Bourbon restoration

He dies in Paris in **1827** and his brain is put on display in a roving anatomical museum in Britain

FROM ALGEBRA TO ASTRONOMY

∫ Laplace also carried out important research into the **theory of determinants in algebra** and devised a **formula to approximate integrals (Laplace's method)**.

He developed **"Laplace transforms"** as a way to **convert complex differential equations into an abstract, yet solvable form**, and **then convert the solutions back to the real world**.

In the 1770s, he used the laws of **gravity, rotation,** and **fluid dynamics** to explain the **Earth's tides**.

Laplace even suggested that there exist **stars with gravity so great they cannot emit any light**—what we would call a **black hole**.

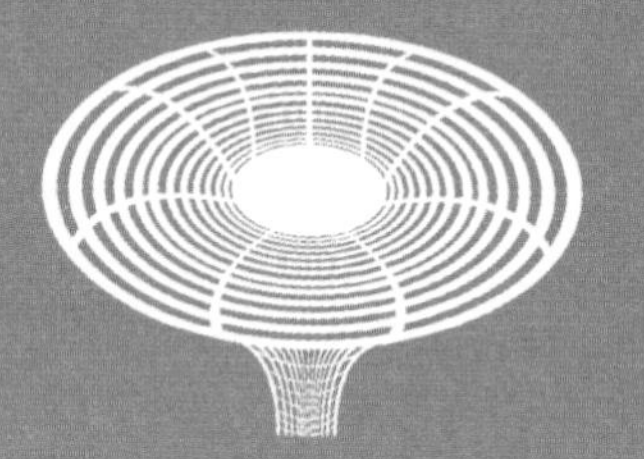

PROBABILITY

In Laplace's seminal work on probability ***Théorie Analytique des Probabilités,*** he presented his tools for **predicting the probability of events occurring in nature**.

Laplace helped to popularize what we now know as **Bayesian statistics**. He also developed the **method of least squares**, and provided one of the first proofs of the **central limit theorem**.

SOPHIE GERMAIN

Barred because of her gender from university, Sophie Germain had to pose as a man to let her mathematical genius shine.

Marie-Sophie Germain is born in Paris in 1776. She has **no formal education** and teaches herself **Greek**, **Latin**, and **mathematics**

Forced indoors for safety during the **French Revolution** in 1789, Germain pores over the work of **Sir Isaac Newton**, **Leonhard Euler**, and **Archimedes**

In 1794, the **École Polytechnique** opens, but being a woman, Germain is barred from entry. However, **lecture notes** could be requested by members of the public

Germain sent her observations to the eminent mathematician **Joseph Louis Lagrange**, under the **pseudonym Monsieur Le Blanc**. Lagrange spots **remarkable talent** and offers to meet. Germain has to reveal her **true identity**, but this doesn't matter to Lagrange, who becomes **her mentor** after 1794

She **dies** in 1831 from breast cancer

ACHIEVEMENTS

In 1816, Germain won a prize offered by the **French Academy of Sciences** to provide a theory to explain the **behavior of vibrating metal plates**. Her study of elasticity won on the third attempt, making her the **first woman to win a prestigious Academy prize**.

When the **Eiffel Tower** was built, engineers had to pay careful attention to the **elasticity of the materials used**. Despite the importance of Germain's theories in this field, she was **not among seventy-two scientists whose names were inscribed on the tower**.

Germain carried out some of her finest work in the field of **number theory**. In particular, she was fascinated by **Fermat's Last Theorem**. Germain devised her own theorem governing solutions to the equation $x^p + y^p = z^p$.

JOSEPH FOURIER

Renowned for discovering the greenhouse effect, Fourier was also a talented mathematician. He survived imprisonment during the French Revolution to become one of France's leading scientists.

NINETEENTH CENTURY

- **1768** Fourier is born in Auxerre in the Burgundy region of France
- **c.1777** He is educated at the local military school by Benedictine monks
- **1795** Fourier teaches mathematics at the École Normale and later at the École Polytechnique
- **1798** He accompanies Napoleon to Egypt as a scientific adviser
- **1807** Fourier presents early work on heat propagation in solid bodies
- **1822** He is appointed a leading member of the French Academy of Sciences
- **1830** Fourier dies after suffering a heart condition that dated back to his time in Egypt

KEY ACHIEVEMENTS

The heat equation: Fourier spent fifteen years studying **heat conduction** and presented his findings in 1822 in the work ***Théorie Analytique de la Chaleur.*** He found that the **flow of heat between adjacent molecules** is **proportional to the tiny difference in their temperatures**.

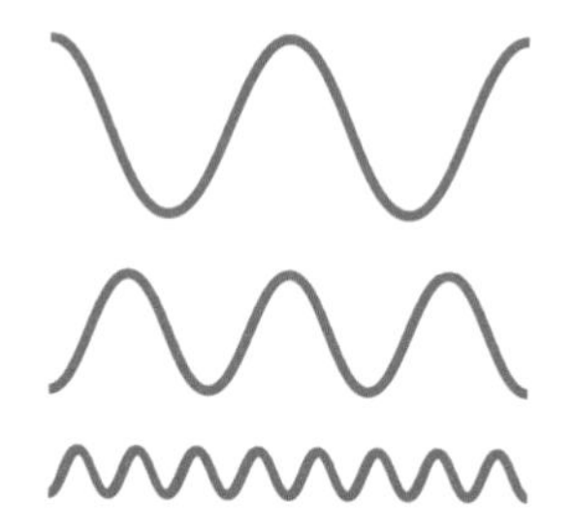

Fourier series: periodic functions can be written as a series of sines and cosines.

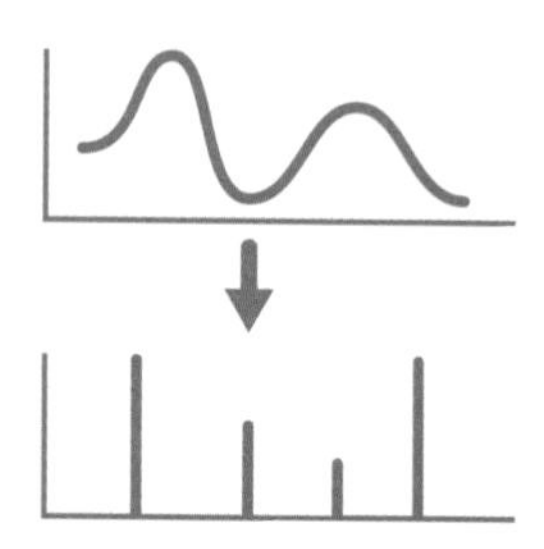

Fourier transform: this extends the series idea to represent **arbitrary functions as a complex exponential integral**.

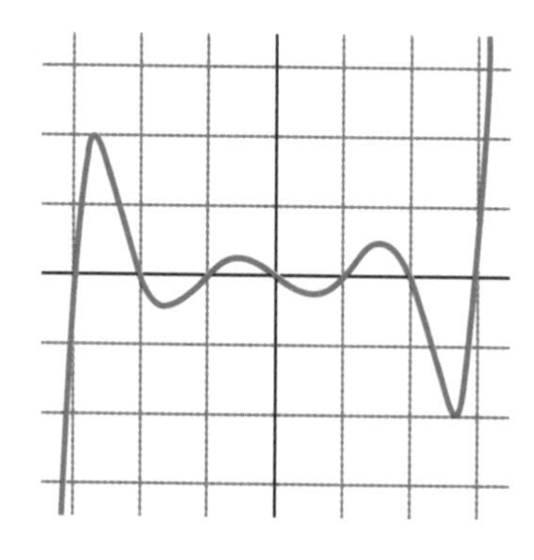

Theorem on polynomial real roots: a technique for finding the number of real (i.e., non-imaginary) roots of a **polynomial** (an equation featuring different integer powers of the unknown variable).

THE GREENHOUSE EFFECT

In the 1820s, Fourier calculated that given Earth's size and distance from the Sun, it **should be much colder** than it is. He was the first scientist to deduce correctly that **Earth's atmosphere must work as an insulator**.

FOURIER ANALYSIS

Fourier analysis is a way to break arbitrary mathematical functions down into a sum of waveforms. It has applications in signal processing, calculus, heat transfer, and many others fields.

WAVE FUNCTIONS

The underlying principle is to **decompose functions** in terms of the **trigonometric ratios sine and cosine**. These can be thought of as **"pure" waves**.

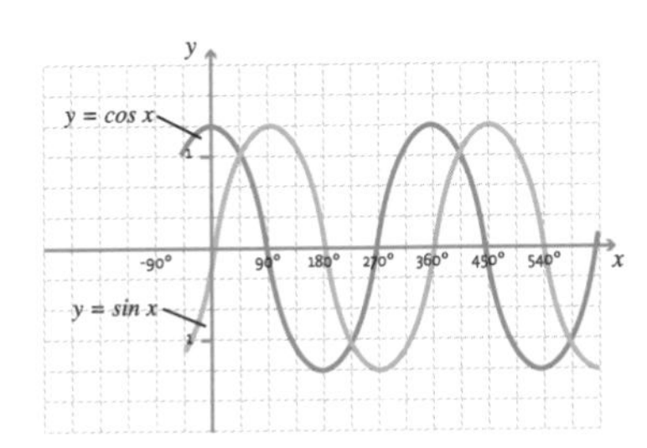

SUPERPOSITIONS

Sine and cosine waves of different **wavelength** (the distance between successive peaks or troughs of the wave) and **amplitude** (the height of the wave) can be **added together to create more complex curves**.

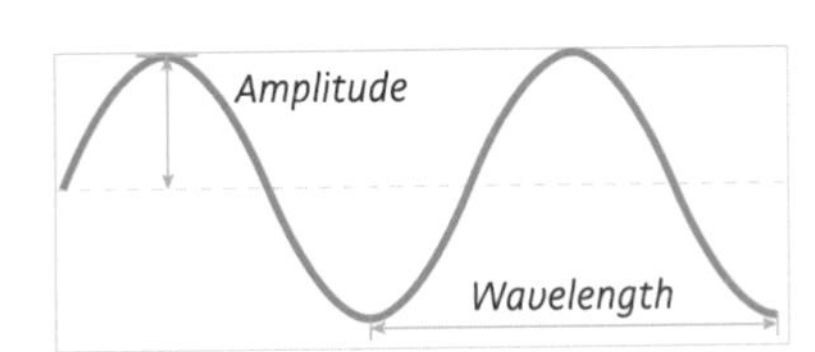

For example, the two **input waves** in the diagram below have the **same amplitude** but **different wavelengths**. They add together to create the **upper waveform**.

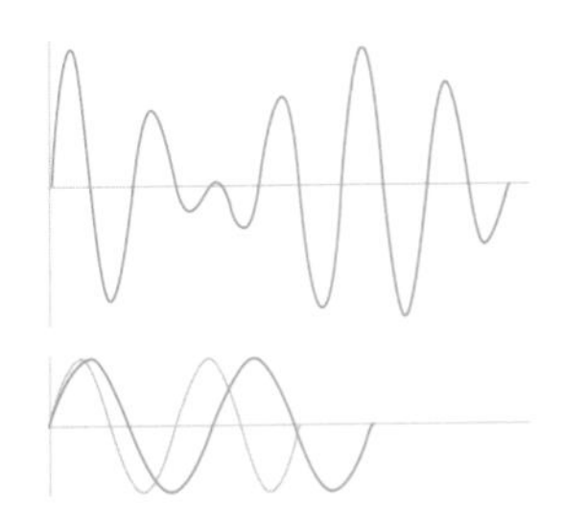

FOURIER'S THEOREM

In 1807, the French mathematician **Joseph Fourier** went further to suggest that **any periodic mathematical function** can be **broken down** into a **series of sines and cosines**. That is,

$$f(x) = \sum_n a_n \sin nx + \sum b_n \cos nx$$

So at any point, x, the value of f is given by a **sum of sine and cosine waves of different wavelengths** (controlled by n), each with **amplitudes** a_n and b_n respectively. This is Fourier's theorem.

Fourier also showed that the **coefficients** an and bn can be calculated for each n, as **integrals** involving f:

$$a_n = \frac{1}{\pi}\int_{-\pi}^{\pi} f(t) \sin nt \, dt$$

$$b_n = \frac{1}{\pi}\int_{-\pi}^{\pi} f(t) \cos nt \, dt$$

EXAMPLE

A **sawtooth waveform** can be expressed as a **sum of sine waves:**

$$y = \sin x + \frac{1}{2} \sin 2x + \frac{1}{3} \sin 3x + \ldots + \frac{1}{n} \sin nx$$

FOURIER TRANSFORMS

Fourier series are a discrete version of Fourier analysis—a *sum* over a particular range of n. The continuous analog of this is known as a Fourier transform and involves replacing the sum with an integral:

$$f(x) = \int_{-\infty}^{\infty} g(t) e^{itx} dt$$

Where $g(t)$, the analogue of the coefficients, is given by

$$g(t) = \frac{1}{2\pi} \int_{-\infty}^{\infty} f(x) e^{-itx} dx$$

Here, e is **Euler's number** (= 2.718...) and i is the **unit imaginary number** ($i^2 = -1$).

MODULAR ARITHMETIC

What if numbers behaved like a wraparound video-game screen, where counting up to a particular number takes you right back to zero? Welcome to the world of modular arithmetic.

THE TWELVE-HOUR CLOCK

An example of modular arithmetic is the **twelve-hour clock**. Adding time to any particular clock read-out is essentially *modulo 12* arithmetic. So, for instance, five hours on from 10:00 is 03:00 (rather than 15:00)—because, on reaching twelve, the **counter resets to zero**.

NOTATION

In general, if the **difference between two numbers**, *a* and *b*, **is exactly divisible by a third number** *m* then *a* and *b* are said to be ***congruent modulo*** *m*.

Formally, a mathematician would write $a \equiv b \ (mod\ m)$.
For example, $8 \equiv 4 \ (mod\ 2)$, because the difference between 8 and 4 divides exactly by 2.
We might write the clock example as $10 + 5 = 15 \equiv 3 \ (mod\ 12)$.

Here are some more:

$$64 \equiv 10 \ (mod\ 6)$$

$$-9 \equiv 6 \ (mod\ 5)$$

$$-11 \equiv -9 \ (mod\ 2)$$

APPLICATIONS

The applications of modular arithmetic are manifold. They include:

Computer programming: used for implementing **cyclic operations** when **looping through data**.

Cryptography: modular arithmetic is used to generate the keys used in public **key cryptosystems**.

Week: days of the week operate according to **modulo 7 arithmetic**.

Banking: international account numbers obey **mod 97 arithmetic for error checking**.

Mathematics: it crops up in various other fields, such as **number theory**, **group theory**, and **knot theory**.

MODDED HISTORY

Modular arithmetic was put forward in its current form in 1801 by **Carl Friedrich Gauss** in his book ***Disquisitiones Arithmeticae.*** Contributions were also made by other mathematicians, including **Euler** and **Lagrange**.

CORRELATION AND REGRESSION

One purpose of mathematical statistics is to unpick the relationships between random variables. Correlation and regression are two powerful tools for achieving this.

CORRELATION

A term introduced by English polymath **Francis Galton**, in the late nineteenth century, correlation is a property of two variables, determining **how variation in one corresponds to variation in the other**.

English statistician **Karl Pearson** developed the **"correlation coefficient,"** a numerical measure of correlation that ranges from 1 **(perfectly correlated)** through 0 **(no correlation)** down to -1 (**anti-correlated**—i.e., increase in one coincides with a decrease in the other).

EXAMPLE

In the data below, x and y have a correlation coefficient of 0.92, and are thus highly correlated.

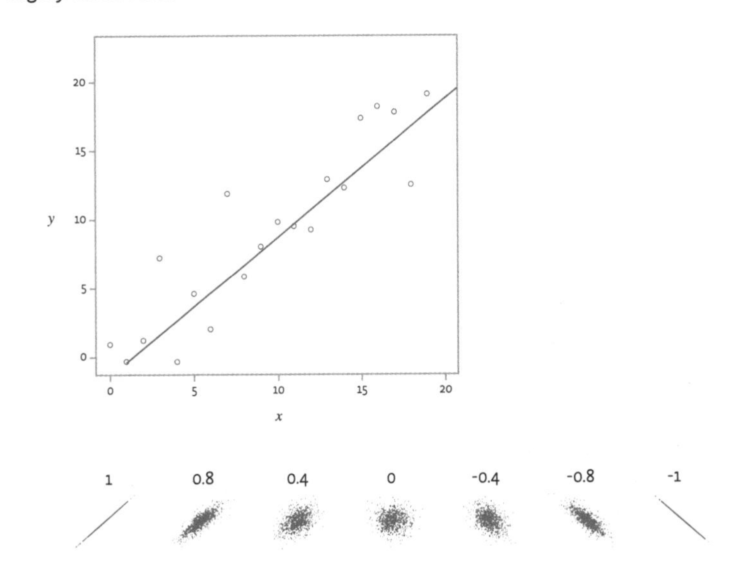

DID YOU KNOW?

To quote the old adage, **"correlation does not imply causation."** For example, the grass being wet and there being a rainbow in the sky are correlated—not because one causes the other but because they both have a **common cause** (i.e., rain).

REGRESSION

Regression is a way to try and **reconstruct the mathematical relationships between variables**.

For example, the line plotted in the example above is calculated using regression and represents the **"best fit"** between the **data points**.

It's governed by the equation

$$y = 1.0125x - 0.4992$$

Here, *y* is called the **"independent variable"** and *x* is the **"explanatory variable."**

Regression **can be applied to more than one explanatory variable**, and used to fit **more elaborate relationships than just straight lines**.

Fitting involves choosing the line that minimizes the sum squared vertical distance between each data point and the line—known as a *least squares fit*.

THE NORMAL DISTRIBUTION

Discovered by Abraham de Moivre more than 200 years ago, the normal distribution is perhaps the most important probability distribution in the whole of statistics.

WHAT IS IT?

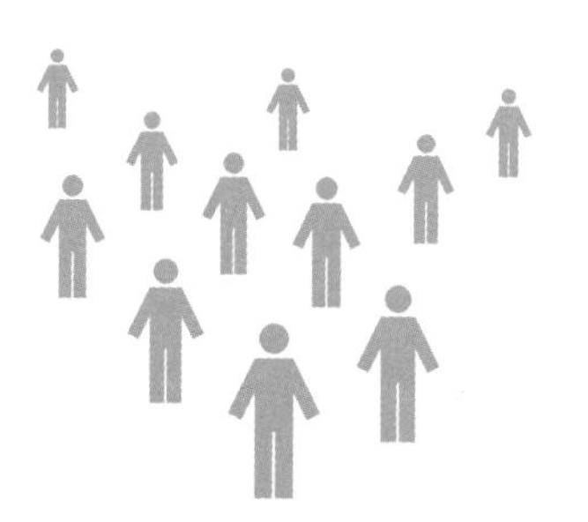

Take a randomly selected bunch of people and measure their heights. Now plot the number of people with each possible height on a **histogram** and you'll end up with a **graph that looks like the normal distribution**. The normal distribution is also known informally as the **"bell curve,"** due to its shape.

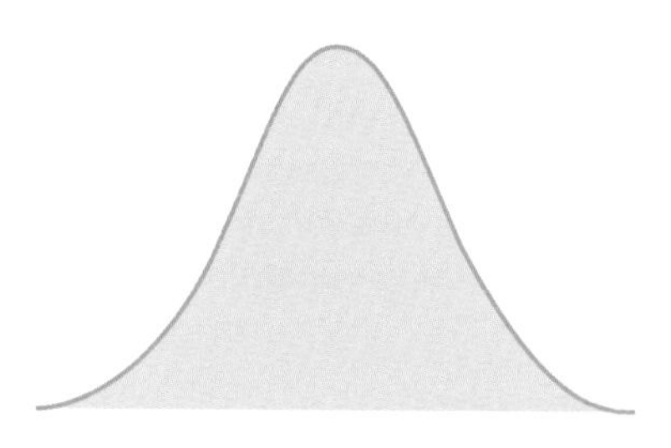

WHAT DOES IT MEAN?

Like other **probability distributions**, the curve indicates the **probabilities of particular random values occurring**—in this case, heights of people, but the distribution can apply in many different situations.

The **peak of the distribution** corresponds to the **average**, or **"mean" value**, often denoted as μ.

The **spread of the distribution** is captured by its **"standard deviation,"** σ, calculated such that 68 percent of the data recorded falls within 1 standard deviation; 95 percent within 2; and 99.7 percent within 3.

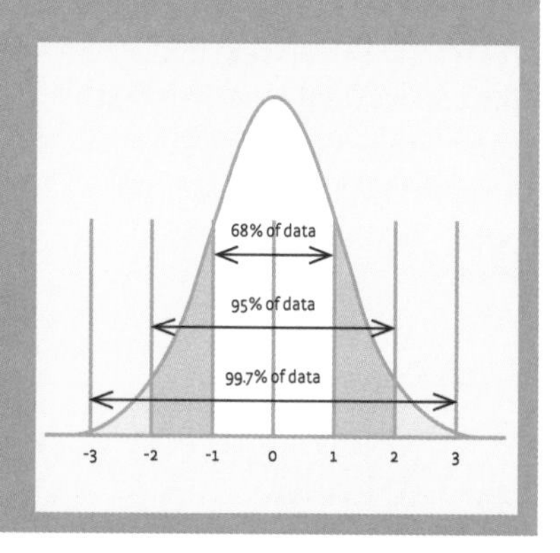

DID YOU KNOW?

The **shape of the curve** is given by the equation

$$y = \frac{1}{\sqrt{2\pi\sigma^2}} e^{-\frac{(x-\mu)^2}{2\sigma^2}}$$

This was discovered by **Carl Friedrich Gauss**, hence the alternative term, **"Gaussian" distribution**.

CENTRAL LIMIT THEOREM

If you add together many random variables, the **central limit theorem** says that their sum will be normally distributed.

So, if you roll a die 1,000 times then, **regardless of the distribution of each roll**, the sum of all 1,000 rolls will obey a **normal distribution**.

THE GOLDEN MEAN

From the Parthenon to the pyramids, the golden mean, also known as the "golden ratio," is everywhere. It ranks alongside π and e as one of the most important constants in mathematics.

WHAT IS IT?

The term "golden ratio" was first coined by the German scientist **Martin Ohm** and is usually abbreviated by the Greek letter φ (phi).

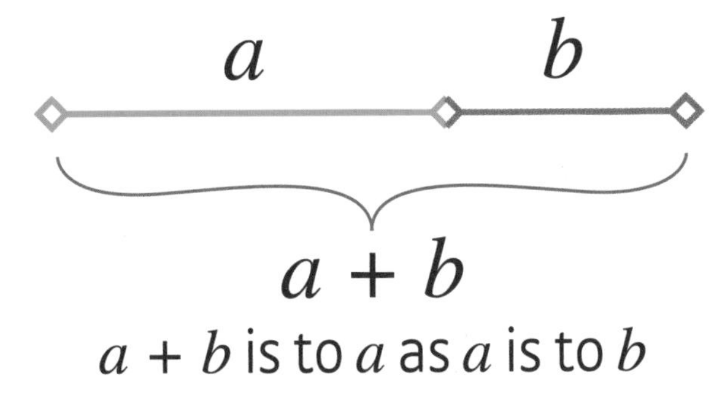

$a + b$ is to a as a is to b

Take a line and divide it into **two segments of differing lengths**, *a* and *b*. They are in the golden ratio if ...

$$a / b = (a + b) / a = \varphi$$

Rearranging this equation gives

$$\varphi = 1 + \frac{1}{\varphi}, \text{ or } \varphi^2 - \varphi - 1 = 0$$

a **quadratic equation** with the solution

$$\varphi = \frac{1 + \sqrt{5}}{2} \approx 1.6180339887498948420$$

NINETEENTH CENTURY

GOLDEN RECTANGLE

Phi has myriad uses in **art**, **architecture**, and **design**, often based around the "golden rectangle." This is a rectangle with side lengths in proportion φ. A golden rectangle with sides *a* and *b* (where *a* is the longer side) has the property that joining it to a square of side length *a* creates a **new golden rectangle** with side lengths *a* and *a* + *b*.

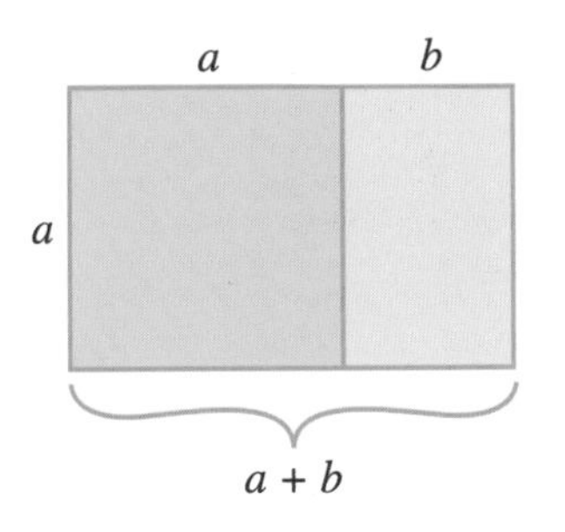

DID YOU KNOW?

There is a very **close relationship** between the **golden mean** and the **Fibonacci sequence**.

0, 1, 1, 2, 3, 5, 8, 13, 21, 34, 55, 89, 144, 233...

Take any number in the sequence and divide it by its predecessor. The **result gets closer to the golden mean as the numbers get bigger:**

5/3 = 1.666666666

8/5 = 1.6

CARL FRIEDRICH GAUSS

Alongside Newton and Archimedes, Gauss ranks as one of history's greatest mathematicians. He made huge contributions to algebra, number theory, and geometry.

KEY ACHIEVEMENTS

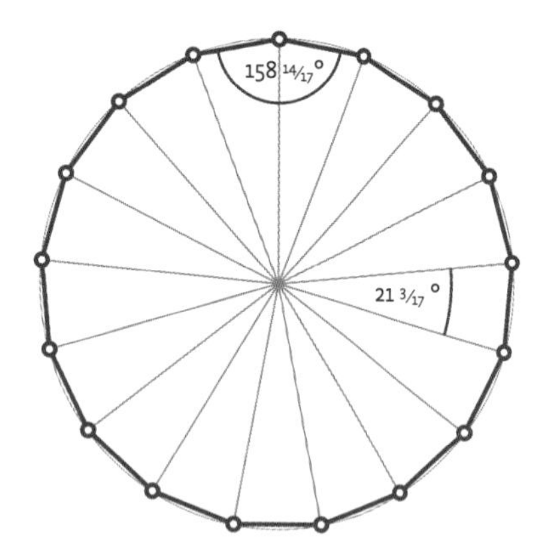

Gauss showed that a regular **polygon** of seventeen sides (a **heptadecagon**) could be **created by using a ruler and compasses**.

He showed this could be done for **any polygon** with a **number of sides equal to a Fermat prime** ($2^{2^n} + 1$, where $n = 0, 1, 2, \ldots$). This was important because it proved a **link between algebra and geometry**—something that mathematicians had been trying to do for around 2,000 years.

In his doctoral thesis, Gauss proved the **fundamental theorem of algebra**—that **every algebraic equation has at least one root or solution**. Mathematicians had grappled with this for centuries.

In **number theory**, he provided the **first two proofs of the law of quadratic reciprocity**, a theorem from the field of **modular arithmetic**.

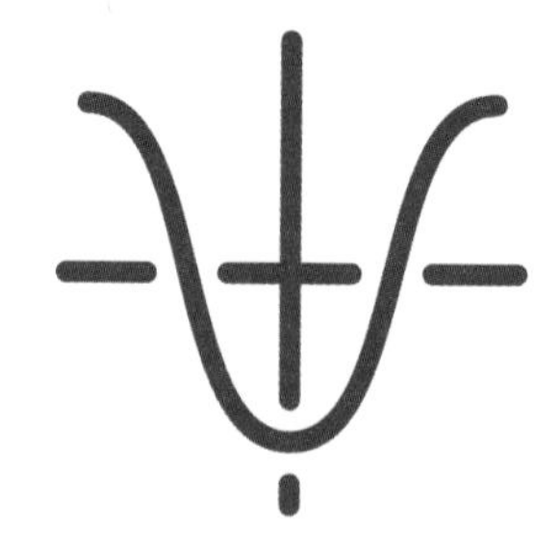

Gauss established the equation underpinning the **"normal distribution,"** an important tool in **statistics** and **probability theory**.

He introduced the **symbol** ≡ to denote **congruence in modular arithmetic**.

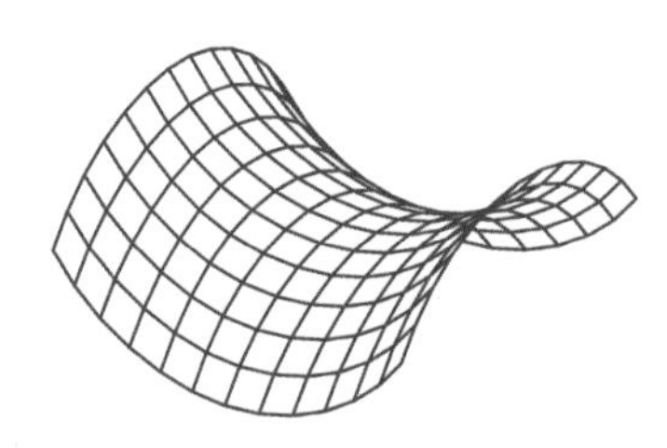

Gauss helped lay the foundations for a **mathematical treatment of non-Euclidean geometries** (curved spaces), realizing the huge potential of this field.

A YOUNG GENIUS

Gauss was born into a working-class family in Brunswick (part of modern Germany) in 1777. He was a **child prodigy** who **could correct arithmetic errors by the age of three**. The **Duke of Brunswick** recognized his early talent and **sponsored Gauss's studies at the University of Göttingen**, where he later became a professor.

Most of his **ground-breaking theories** were **formed as a teenager** and published in the seminal work ***Disquisitiones Arithmeticae*** (Arithmetical Investigations, 1798).

He **died** in Göttingen in 1855.

COMPUTING

Computers are one of the ultimate practical applications of mathematics and logic. And today, they are in everything from your phone to your refrigerator.

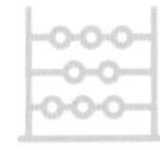

2500 BC The ancient Sumerians build an early hand-operated calculating aid—the abacus

200 BC Indian mathematician Pingala develops binary numbers, allowing a value to be represented by the on–off state of a switch

125 BC The Greek "Antikythera mechanism," thought to be one of the earliest mechanical computers, is used for charting the positions of the planets

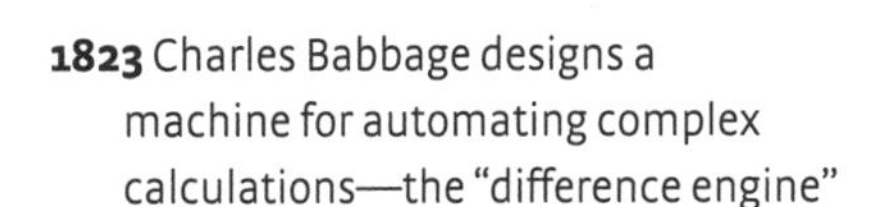

1823 Charles Babbage designs a machine for automating complex calculations—the "difference engine"

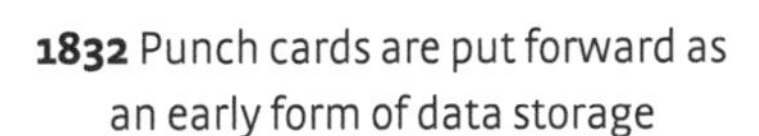

1832 Punch cards are put forward as an early form of data storage

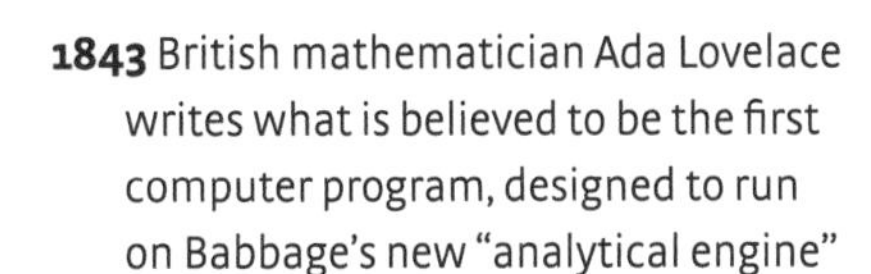

1843 British mathematician Ada Lovelace writes what is believed to be the first computer program, designed to run on Babbage's new "analytical engine"

1847 George Boole develops his algebra for manipulating true/false logical variables

1936 Alan Turing publishes a scientific paper in which he proposes a mathematical model of a general-purpose computer, now known as a "Turing machine"

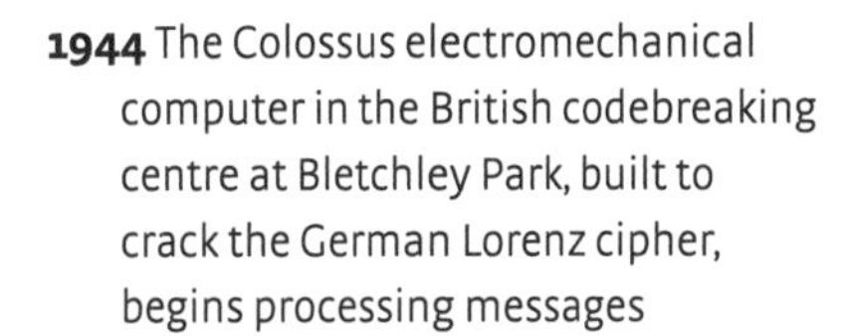

1944 The Colossus electromechanical computer in the British codebreaking centre at Bletchley Park, built to crack the German Lorenz cipher, begins processing messages

1946 ENIAC, a general-purpose, fully electronic computer, is unveiled at the U.S. Ballistic Research Laboratory, in Maryland

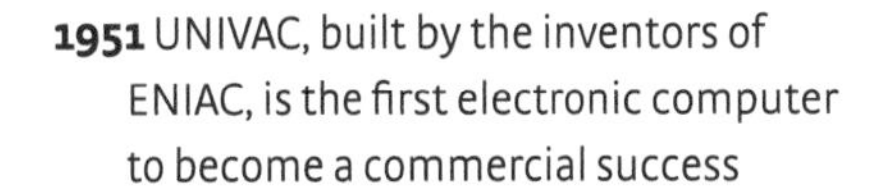

1951 UNIVAC, built by the inventors of ENIAC, is the first electronic computer to become a commercial success

1953 The University of Manchester, England, builds the first transistor-based computer

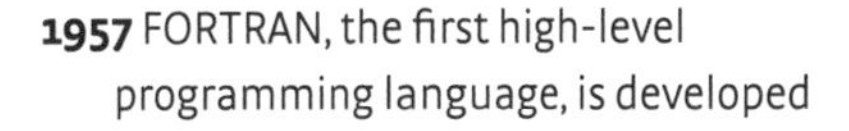

1957 FORTRAN, the first high-level programming language, is developed

1971 The first commercially available microprocessor, packing huge numbers of transistors onto a semiconductor chip, is produced

And from here, computers migrated out of the lab and into every aspect of our lives

AUGUSTIN-LOUIS CAUCHY

Cauchy penned some 800 research papers, laying the foundation for modern-day mathematical analysis, and making significant contributions to mathematical physics.

1789 Born at the outset of the French Revolution, Cauchy is the son of a high-ranking police officer

1793 During the Reign of Terror, the family has to flee Paris

1802 On their return, Cauchy excels at the École Centrale du Panthéon, one of Paris's finest secondary schools

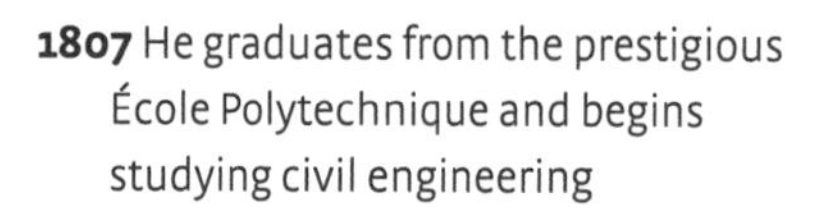

1807 He graduates from the prestigious École Polytechnique and begins studying civil engineering

1812 After working on Napoleon's naval base at Cherbourg, Cauchy returns to Paris dreaming of an academic career in mathematics

1815 Cauchy's brilliance wins him the role of professor of mathematics at the École Polytechnique, which he retains after Napoleon's fall

1818 Cauchy marries Aloïse de Bure, whose family firm published many of his treatises. Cauchy's life work fills twenty-seven volumes

1830 After the July Revolution, a disenfranchised Cauchy goes into exile for eight years

1838 He returns to Paris and is appointed to a number of senior roles at colleges, including the Academy of Sciences

1857 He dies in Sceaux, Paris, of a bronchial condition. His name is one of the seventy-two scientists inscribed on the Eiffel Tower

KEY ACHIEVEMENTS

Cauchy won the **Grand Prix** of the **French Academy of Sciences** in 1816 for his **theories on wave propagation**.

Cauchy was the **first person to prove the Fermat polygonal number theorem**.

This states that each positive integer number can be written as a sum of no more than *n* *n*-gonal numbers. Where, for example, the 3-gonal numbers are just the triangular numbers—1, 3, 6, 10, ..."

$\int$ He developed **complex function theory**, producing what has become known as Cauchy's integral theorem.

In 1827, Cauchy found a way to describe the detailed way in which **materials deform under stress**.

NON-EUCLIDEAN GEOMETRY

For millennia, the rules of geometry were governed by the axioms of Euclid. Then, in the nineteenth century, mathematicians began to wonder what might happen if they relaxed these constraints.

- **300 BC** Euclid's defining treatise on flat-space geometry, *Elements*, is written
- **c.1830** Russian Nikolai Lobachevsky and Hungarian János Bolyai independently publish their work on hyperbolic geometry
- **1854** German mathematician Bernhard Riemann introduces the mathematical tools for studying curved spaces, Riemannian geometry
- **1859** British mathematician Arthur Cayley begins the study of elliptic geometry
- **1868** Italian mathematician Eugenio Beltrami devises Poincaré's disk

PARKING PARALLEL

Perhaps the most important of all Euclid's axioms was the **parallel postulate**—the notion that **parallel lines should remain parallel forever**.

In the early nineteenth century, a number of researchers conceived of a geometry where **parallel lines diverge**.

Whereas Euclid's rules concerned flat space, the new approach became known as **"hyperbolic geometry,"** because it described **space** that was **curved into a shape resembling a hyperbola**.

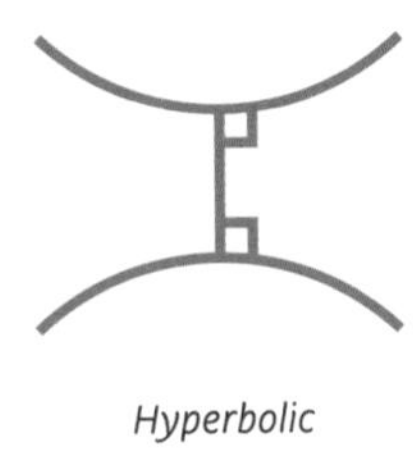

Hyperbolic

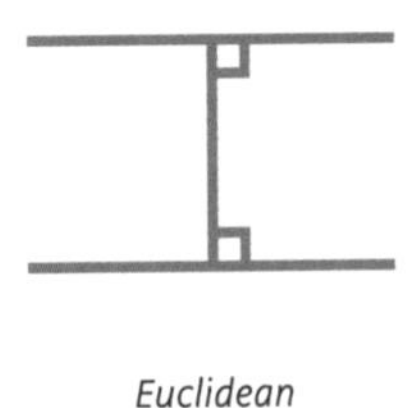

Euclidean

Elliptic

HIGHER FUNCTIONS

There also exist **hyperbolic equivalents of flat-space functions**. So whereas **flat-space trigonometry**, for example, is based on sin and cos, these are replaced by **analogous functions** in **hyperbolic space**, called **sinh** and **cosh** (pronounced "shine" and "cosh").

ELLIPTIC GEOMETRY

Hyperbolic geometry can be thought of as **negatively curved**, but there **also exist spaces** with **positive curvature where parallel lines converge**. These are **elliptic geometries**. The **simplest is a sphere like the Earth**. Imagine two lines at the equator pointing north—these are **parallel**, and yet they **cross at the north pole**.

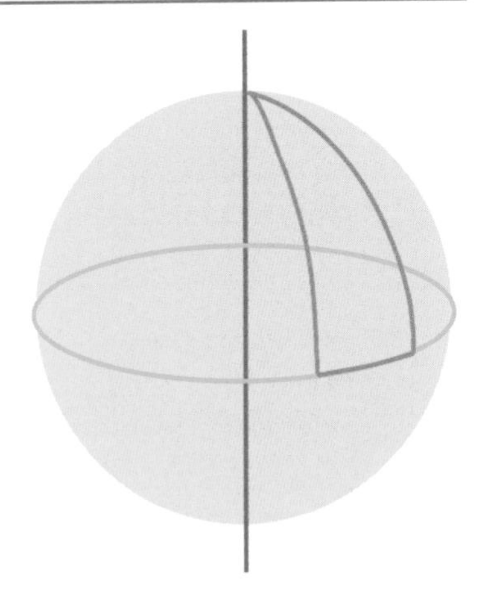

GROUP THEORY

Group theory is used for classifying mathematical objects according to their properties and symmetries. It has found applications in particle physics and cryptography.

NINETEENTH CENTURY

WHAT IS IT?

Group theory as a **branch of mathematics** was **first developed** in the early nineteenth century by the fiery young French mathematician, **Évariste Galois**.

ÉVARISTE GALOIS

- **Born** October 25, 1811, in Bourg-la-Reine in the French Empire
- **Graduates** in mathematics from the **École Normale, Paris**, on December 29, 1829
- **Publishes** three mathematics papers in the year 1830
- **Arrested** and **imprisoned** following a **political protest** in 1831. **Develops his mathematical work while incarcerated**
- **Dies** May 30, 1832, at age twenty, from a gunshot wound sustained in a **duel**

He posited that the **members of a group, *G*, must satisfy the following four axioms:**

1. The **result of combining two elements of *G* must also be an element of *G***.
2. Every group has a **unique *identity*** element, which **has no effect when combined with any other**.
3. **For every element in G, there is an inverse** that **combines with it to give the identity element**.
4. Combining elements within *G* is an **associative process;** that is, ***x* combined with (*y* combined with *z*) is the same as (*x* combined with *y*) combined with *z***.

EXAMPLE

Integer numbers can be thought of as a **group**, the **elements of which are combined through addition**.

The identity element is 0, since for any element, x, $x + 0 = x$.

The inverse of x is $-x$, since $x + -x = 0$, the **identity element**.

And $x + (y + z) = (x + y) + z$, so they are **associative**.

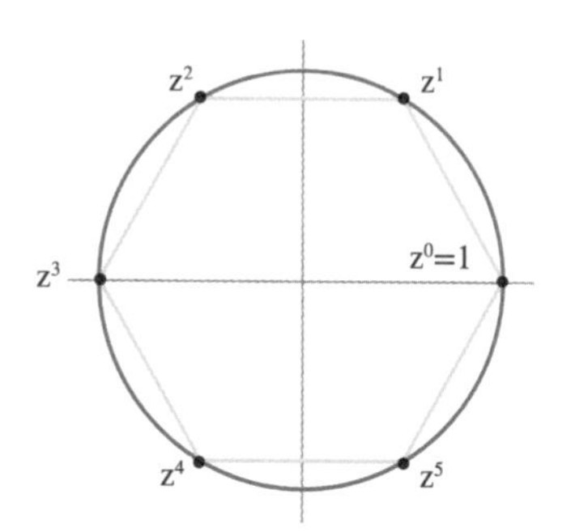

APPLICATIONS

Group theory has been used to great effect to **classify symmetries in the laws of fundamental physics**. It's also used in the **construction of codes and ciphers**. It can even describe the allowed moves on a **Rubik's cube**.

DIFFERENTIAL GEOMETRY

Once it's accepted that curved, or "non-Euclidean" geometries can exist, the challenge then becomes to describe them mathematically. That's where differential geometry comes in.

GAUSSIAN CURVATURE

Roll up a sheet of paper and it looks curved. However, this **curvature is just "extrinsic"**—one can simply unroll the paper and lay it flat. A sphere, on the other hand, is **"intrinsically" curved**. One can't take its surface and lay it flat **without cutting or tearing it**—a bit like the peel from an orange.

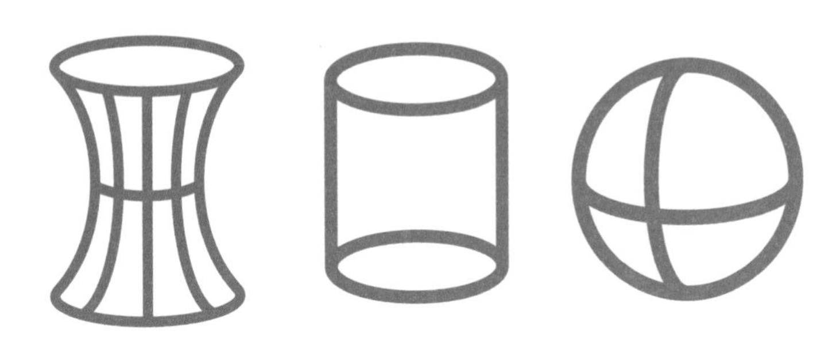

German mathematician **Carl Gauss** realized this distinction in the early nineteenth century, and called it his **"Theorema Egregium."** He used this to produce a measure of how **intrinsically curved a space** is, called **"Gaussian curvature."**

It is **negative for hyperbolic spaces, zero for spaces that are flat or cylindrical**, and **positive for elliptical or spherical spaces**.

GEODESICS

In **Euclidean flat space**, the **shortest route between two points is a straight line**. In **curved space** these **shortest paths, called "geodesics," also become curved**. For example, on a sphere such as the **Earth**, the geodesics are **"great circles"** encompassing the globe.

RIEMANNIAN MANIFOLDS

German mathematician **Bernhard Riemann** took the idea further, developing a **mathematical framework by which to measure distances and angles**, and **other geometrical properties**, on a **general curved surface**.

In fact, **not all curved surfaces admit themselves to Riemann's treatment**. Those that do, called **"Riemannian manifolds,"** are **smooth** so that **calculus** can be performed on them everywhere.

GENERAL RELATIVITY

In 1915, **Albert Einstein** applied the principles of **differential geometry** to **space** and **time**. His **general theory of relativity** posited that **gravity manifests itself by curving space-time**. Einstein developed an **equation linking the geometrical curvature of space-time to the material content of the universe**.

$$R_{\mu\nu} - \frac{1}{2} R g_{\mu\nu} = (8\pi G)\, T_{\mu\nu}$$

Curvature of space-time — *Constants* — *Energy and momentum*

CHARLES BABBAGE

History was made by Babbage, who invented the world's first computers. These mechanical devices formed the basis of modern-day computer technology.

He was born in London on **December 26, 1791**, the son of a banker.	In **1810**, he went up to Cambridge University.	Babbage and his friends established the Royal Astronomical Society in **1820**.	He was Lucasian Professor of Mathematics at Cambridge University for **eleven** years.	Babbage died in **1871**, after a forty-year career. He refused a knighthood and a baronetcy.

During the early nineteenth century, calculations were carried out **manually** using **printed mathematical tables**. Babbage recognized how limiting this process was

Babbage was convinced that a **machine could be built to automate calculations quickly and accurately**. In 1823, he received **government support** for his projected design: the "**difference engine**"

The device consisted of **toothed wheels** that operated in a **clockwork fashion** and was powered by a **crank handle**. It had a **twenty-digit capacity** and the **output** was **stamped onto soft metal** that could be turned into **printing plates**

A small working model was produced in 1832, which **could raise numbers to second and third powers** and **extract the roots of quadratic equations**. Despite government funding, **costs ran out of control**, and the **full-scale difference engine was never completed**

In the early 1830s, Babbage began work on a **larger, more sophisticated machine**, designed to **calculate any mathematical function**. This **"analytical engine"** took its **input from punched cards**, and could **store and process 1,000 forty-digit numbers**. Again, sadly, he **never completed it**

A **working difference engine was later constructed** to Babbage's design at **London's Science Museum** in 1991

PETER GUSTAV LEJEUNE DIRICHLET

Inspired by the work of Carl Friedrich Gauss, Dirichlet was a mathematician who made contributions to number theory, Fourier analysis, differential equations, and statistics.

Dirichlet was born in 1805 in Düren, a town on the Rhine, in Germany, which was under French control.

He was the youngest of seven children in a family of modest means. His parents supported his education, and he studied at the **Jesuit Gymnasium** (Catholic school) in **Cologne**, under the brilliant physicist and mathematician Georg Ohm.

Dirichlet studied mathematics in Paris and became a **private tutor**. He returned to Germany, where he became **professor of mathematics at the universities of Breslau**, **Berlin**, and **Göttingen**.

Dirichlet **died** in Göttingen in 1859.

FAMOUS FOR

Number theory was Dirichlet's main fascination and his proof of **Fermat's Last Theorem** (below) for the case n = 5 brought him instant fame. He later completed a proof for the case n = 14.

$$x^n + y^n = z^n$$

In 1837, he published ***Dirichlet's Theorem on Arithmetic Progressions,*** which deployed **mathematical analysis** to tackle **algebraic problems**. In the process, he created his own branch of **analytic number theory**.

He proposed the modern concept of a **function** (a **rule or law that defines relationships between variables**). For instance, y = f(x), in which for every x, there is a unique y associated with it.

The **Dirichlet distribution**, from **Bayesian probability theory**, is named after him.

DID YOU KNOW?

Dirichlet has a **Moon crater** and an **asteroid** (11665) **named in his honor**.

His **brain** is preserved in the **physiology department** at the **University of Göttingen**, along with the **brain of Gauss**.

ADA LOVELACE

Lovelace was a Victorian aristocrat who was fascinated with mathematics. At a time when few women were accepted in academic circles, she became the first computer programmer.

December 10, 1815 The Honorable Augusta Ada Byron is born in London—the daughter of Lord Byron and his wife Anne Isabella Milbanke

1816 Two months after her birth, Bryon and his wife separate

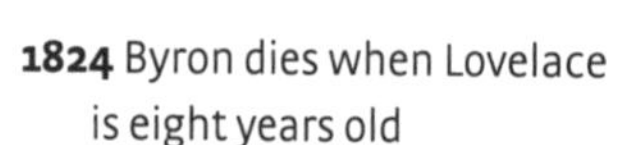

1824 Byron dies when Lovelace is eight years old

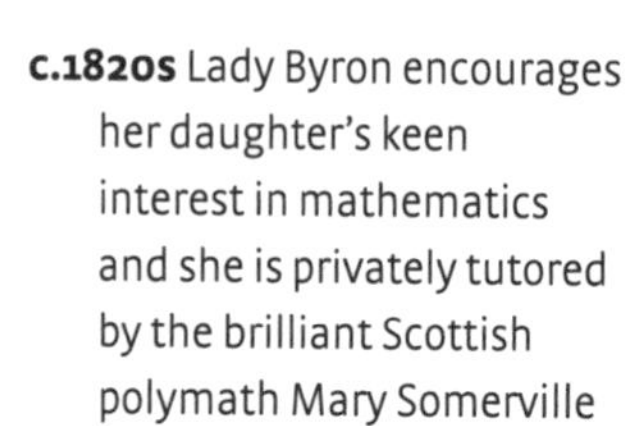

c.1820s Lady Byron encourages her daughter's keen interest in mathematics and she is privately tutored by the brilliant Scottish polymath Mary Somerville

1833 Somerville introduces Lovelace to Charles Babbage, inventor of the mechanical computer, with his "difference engine." Lovelace is fascinated and Babbage is equally impressed by her. He becomes her mentor and lifelong friend

1835 She marries William, 8th Baron King, who later becomes the Earl of Lovelace. Three children follow

1842 Lovelace translates an article by the Italian mathematician Luigi Menabrae on Babbage's new computer, the "analytical machine," his most successful design. She appends elaborate notes, describing how the machine worked—an incredibly complex task

1843 Lovelace's work is published to wide acclaim. In it, she describes how the analytical machine could be programmed to compute the Bernoulli numbers. This is believed to be the first published computer algorithm

November 27, 1852 At the age of thirty-six, Lovelace dies of uterine cancer. The early programming language "Ada" was later named in her honor

WILLIAM ROWAN HAMILTON

Irish mathematician William Rowan Hamilton discovered quaternions, extensions of imaginary numbers into higher dimensions, and reformulated the classical theory of mechanics.

A LIFE IN MATH

The son of a solicitor, Hamilton is born in Dublin, in 1805

At the age of three, he is sent to live with his uncle, who runs a school in County Meath. Hamilton shows **remarkable talent in classical and ancient languages**. He studies **mathematics**, **physics**, and **classics** at **Trinity College, Dublin,** where he is a top student

In 1827, Hamilton is appointed **professor of astronomy at Trinity and Royal Astronomer of Ireland**. He lives at **Dunsink Observatory** all his adult life

Hamilton makes his breakthrough discovery in the **algebra of quaternions** while walking into Dublin one day

He is knighted by the **lord lieutenant of Ireland** in 1835

Hamilton **dies** in 1865, at age sixty, after a severe attack of gout, brought on by overeating and drinking

BRAINWAVE BY THE BRIDGE

Hamilton researched **optics** and **dynamics**, but his greatest achievement lay in the discovery of **quaternion algebra**.

He understood that **complex numbers** could be presented as **points on a plane**—with the **real part on one axis** and the **imaginary bit on the other**. What he wanted to do was place them in **3-D space** and **fathom how they could be multiplied and divided**.

On October 16, 1843, Hamilton had a brainwave while walking into Dublin with his wife. It dawned on him to use **quadruplets instead of triplets**. He couldn't resist **carving his equation for "quaternions" into the stonework of Brougham Bridge**.

$$i^2 = j^2 = k^2 = ijk = -1$$

Hamilton devoted most of his life to their study and **applications in geometry**.

CLASSICAL MECHANICS

In 1833, Hamilton developed a formulation of classical mechanics based around what became known as the Hamiltonian of a physical system: the sum of its kinetic energy and potential energy. It's used today in quantum physics.

QUATERNIONS

Quaternions extend complex numbers into higher dimensions. They have applications in describing rotation in mechanics and computer graphics, and are used widely in theoretical physics.

ARGAND DIAGRAM

Complex numbers consist of a **real part** and an **imaginary part**, generally taking the form $z = a + bi$, where i is the unit imaginary number, defined such that $i^2 = -1$, and a and b are ordinary real numbers.

Complex numbers can be plotted on a **2-D plane**, called an **Argand diagram**, with the **real component on the horizontal axis** and **imaginary component on the vertical**.

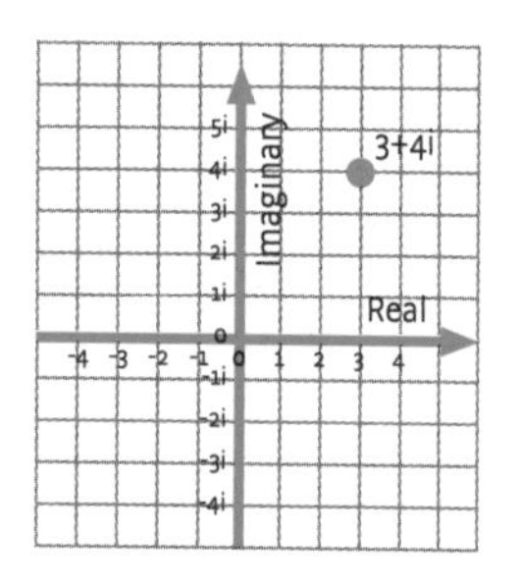

WHAT ARE QUATERNIONS?

Nineteenth-century Irish mathematician **William Rowan Hamilton** wondered whether it was possible to add an **extra imaginary component**, extending the **Argand plane** into **three dimensions**.

However, he found the math only worked by adding **two extra imaginary dimensions**. Hamilton's **new complex numbers** took the general form $z = a + bi + cj + dk$, where a, b, c, and d are real numbers and i, j, and k are distinct imaginary units, obeying the relation:

$$i^2 = j^2 = k^2 = ijk = -1$$

Hamilton called his four-dimensional complex numbers "quaternions."

HOW DO THEY WORK?

The below definition implies that:

$ij = k$, $ij = -k$,
$jk = i$, $jk = -i$,
$ki = j$, $ki = -j$.

Multiplying quaternions **does not commute**—e.g., ij is not the same as ji.

Quaternions describe **rotation**, since the **effect of multiplying a "pure" imaginary j by an i is to rotate the j into the k direction**.

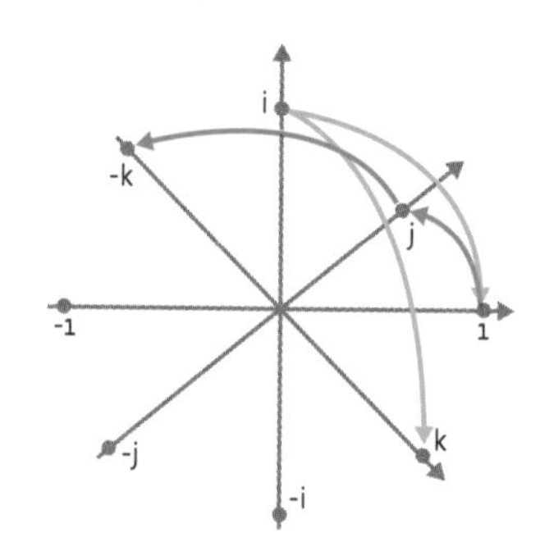

APPLICATIONS

Rotation: the ability of quaternions to describe rotation has led to their use in **computer graphics** and **aircraft navigation**.

Physics: quaternions can be used to formulate **electromagnetism** and are **analogous to the matrices that describe the spin of subatomic particles in quantum theory**.

OCTONIONS

The **octonions** are an **extension of quaternions** from **four to eight dimensions**. They are a **hot topic of research**, with speculation that they **may offer insights to the elusive unification of gravity and quantum physics**.

CATALAN'S CONJECTURE

This theorem in number theory was put forward in 1844 by the Belgian-born mathematician Eugène Charles Catalan. It was finally proven a century and a half later, in 2002.

THE CONJECTURE

Catalan noticed that the **numbers 8 and 9 can be written as 2^3 and 3^2**, respectively. That is, two integers, both raised to integer powers, yield two consecutive integers. His conjecture was that, **with the exception of 0 and 1, 2 and 3 are the only powers for which this is possible**.

The problem can be stated through the **Diophantine equation**,

$$x^a - y^b = 1$$

and amounts to claiming that the **only solution, for *a* and *b* both greater than 1, is**

$$a = 2, b = 3, x = 3, y = 2$$

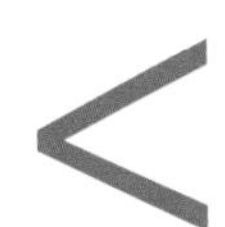

1343 The conjecture is believed to have been known to the French scholar Gersonides

1844 Catalan makes his statement of the problem

1976 Dutch mathematician Robert Tijdeman is able to establish bounds on a, b, x, and y

2002 In April of this year, 158 years after the problem was first posed, Catalan's conjecture is finally proven, by Romanian mathematician Preda Mihăilescu

EUGÈNE CHARLES CATALAN

1814 Catalan is born on May 30 in the city of **Bruges**, the son of a jeweler

1848 With strong **left-wing political views**, Catalan participates in the **People's Spring Revolution**

1865 He becomes chair of **mathematical analysis at the University of Liège**

1894 Catalan **dies** on February 14, in the Belgian city of Liège

TRANSCENDENTAL NUMBERS

If you thought irrational numbers were weird, then transcendental numbers are even stranger, because they cannot be constructed using ordinary algebra.

NINETEENTH CENTURY

1844 Joseph Liouville gives the first examples of transcendental numbers

1873 Charles Hermite shows that *e* is transcendental

1874 Georg Cantor begins his work on the infinite nature of transcendental numbers

1882 Ferdinand von Lindemann demonstrates the transcendental nature of π

1900 David Hilbert poses his "seventh problem"—if *a* is an algebraic number and *b* is irrational but also algebraic, then is a^b always transcendental?

1936 Aleksandr Gelfond and Theodor Schneider prove that the answer to Hilbert's seventh problem is "yes"

WHAT ARE THEY?

The **square root of an integer that's not a perfect square will be an irrational number**. For example,

$$\sqrt{2} = 1.41421356\ldots$$

This is **irrational because it has an infinite number of digits**, but it **can be written as a standard algebraic operation** (in this case, a square root) **applied to a rational number** (in this case, 2).

Transcendental numbers, on the other hand, **cannot be built in this way**. Examples include the **geometric ratio** π and **Euler's number *e***.

They are also known as **"non-algebraic" numbers**.

THE COMMONEST NUMBERS

Despite their seeming weirdness, the **transcendentals are the most common type of number**. In the 1870s, **Georg Cantor** first showed that there exists a **countable infinity of non-transcendental numbers**, and then that the transcendental numbers are ***uncountably* infinite**. That means that **almost every number is, in fact, transcendental**.

EXAMPLES

Known transcendental numbers:

e, π, e^{π}

e^{a} (when a is algebraic and not 0)

a^{b} (where *a* is algebraic but not 0 or 1, and *b* is irrational but not transcendental)

$sin(a)$

$cos(a)$

$tan(a)$

$ln(a)$

The above four examples are transcendental when *a* is algebraic and non-zero. Note, this means that log and trigonometric functions are not algebraic.

GEORG FRIEDRICH BERNHARD RIEMANN

In his short career, this German geometrist made history by laying down the mathematical equations that Albert Einstein used as the framework for his general theory of relativity.

September 17, 1826 Riemann is born in the Kingdom of Hanover. He suffers nervous breakdowns as a child and exhibits social anxiety

10/10

1840 to 1846 Riemann attends the lyceum school in Hanover and the high school Johanneum Lüneburg, where his mathematical genius astonishes teachers

1846 At the University of Göttingen, Carl Friedrich Gauss urges Riemann to study mathematics instead of theology. Riemann takes his advice and attends the University of Berlin, guided by luminaries, such as Peter Dirichlet

1854 He presents his theories on geometry, which would later become the mathematics of general relativity

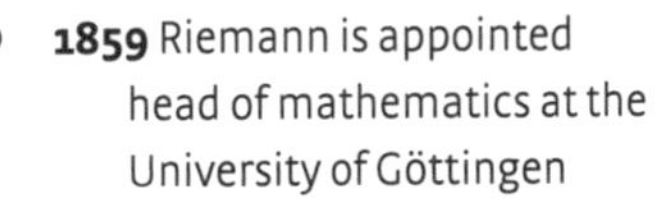

1859 Riemann is appointed head of mathematics at the University of Göttingen

1862 He marries Elise Koch and they have a daughter

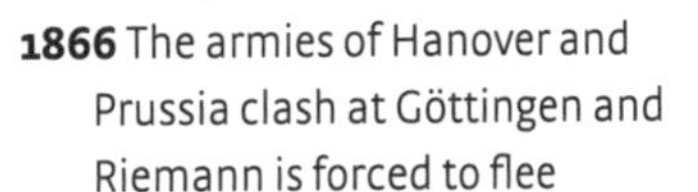

1866 The armies of Hanover and Prussia clash at Göttingen and Riemann is forced to flee

1866 He dies of tuberculosis while staying near Lake Maggiore in Italy

HIS LASTING LEGACY

Riemann **changed the course of science** through his **innovative way of considering geometry**. He believed that the key elements of geometry are **a space of points (called a manifold)** and a system that measures distances along the curves of that space.

He also believed that the space could be of **infinite dimension**. This formed the **mathematical basis of 4-D geometry** that is the **key to understanding space-time in Einstein's theory of relativity**.

NIELS HENRIK ABEL

This brilliant young Norwegian died at the age of twenty-six; however, his discoveries—covering fields such as quintic equations and elliptic functions—left a lasting legacy.

WHO WAS ABEL?

Born in **Nedstrand, Norway,** in 1802, Abel was the son of a poor Lutheran minister. He entered the **Royal Frederick University in Christiania (Oslo)** in 1821, where he was already reputed be the **most knowledgeable mathematician in the country**.

MAJOR BREAKTHROUGHS

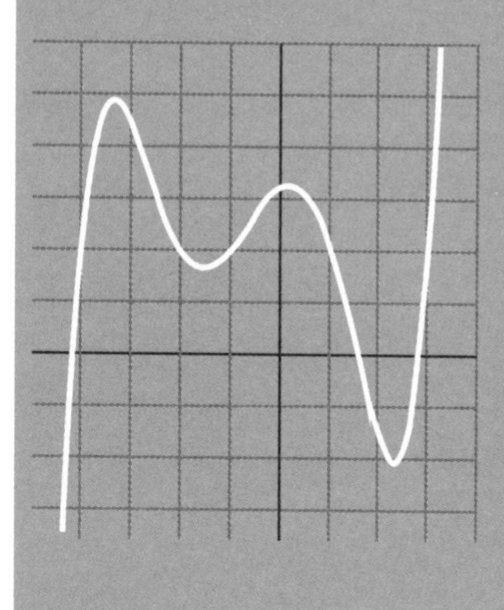

Abel attempted to find a solution to **quintic equations**, which mathematicians had grappled with for 250 years. These are functions defined by a **polynomial of degree 5**. For example,

$$g(x) = ax^5 + bx^4 + cx^3 + dx^2 + ex + f = 0$$

where a, b, c, d, e, and f are **real numbers** and a is **non-zero**. Abel discovered that a **general algebraic solution to a fifth, or higher degree, equation such as this is impossible**.

Abel was the **first mathematician to solve an integral equation**—where the **unknown variable, x, lies inside an integral calculus operation**. For example,

$$\int_0^x \frac{\phi(s)}{\sqrt{x-s}}\, ds = 0$$

In 1826, Abel went to **Paris**, which was a Mecca for mathematicians. He wrote a **key work** on the **theory of integrals of algebraic functions**. This contained what is now known as **Abel's theorem**, which is important in the **theory of elliptic functions and their integrals**.

TRAGIC LOSS

Abel contracted **tuberculosis** in Paris and became very ill on a sledge trip to see his **fiancée Christine Kemp** in Froland. He **died** on April 6, 1829.

THE ABEL PRIZE

Abel's legacy lives on through **the eponymous award** given yearly by the King of Norway to an **outstanding mathematician**. It was established in 2002 (the bicentenary of Abel's birth) and **directly modeled on the Nobel Prize**.

THE FOUR-COLOR THEOREM

How many different colors does it take to shade a 2-D map so that no two adjacent territories are the same color? Experience suggests four. However, proving that was another matter.

1852 Augustus De Morgan poses the four-color conjecture in a letter to William Rowan Hamilton

1879 Arthur Cayley publishes work expanding on the problem and how it might be solved

1890 Percy Heawood finds a problem with a candidate proof of the theorem put forward by Alfred Kempe eleven years earlier

1976 Kenneth Appel and Wolfgang Haken publish their computer proof of the theorem

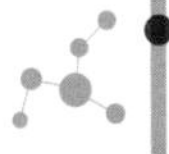

1994 A new algorithm reduces Appel and Haken's 1,938 unique graphs to just 633

WHAT IS IT?

The problem was first suggested to the Irish mathematician **William Rowan Hamilton** in 1852 in a letter from his colleague **Augustus De Morgan**.

In fact, **some simplifications were required. Regions coming together at a point didn't count as adjacent.** And the additional **assumption** was **made that all countries are contiguous**—unlike in the real world where, for example, Alaska is not contiguous with the rest of the United States.

MAPS TO GRAPHS

The first step toward a proof was to find a **systematic way to classify maps**. This was done using **graph theory**. **Distinct regions** of the map were assigned to vertices of the graph, while **edges of the graph represented the borders between each region**.

Even then, there were **9,000 possible graphs** that **could be drawn**.

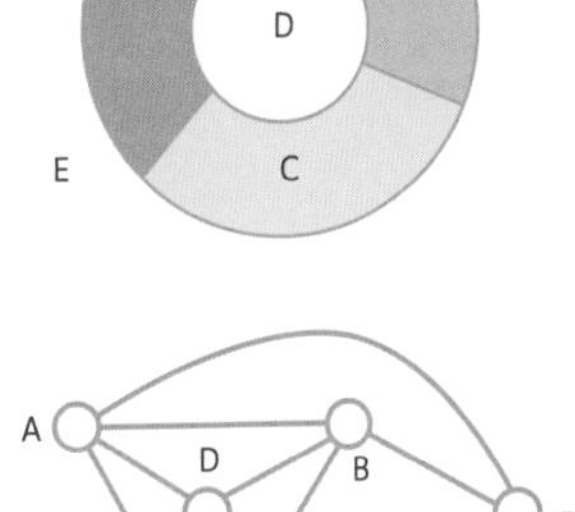

THE PROOF

By 1976, **Kenneth Appel** and **Wolfgang Haken**, of the University of Illinois, were able to **reduce the number of unique configurations to 1,938**. Then they wrote a computer program to check that **none of these required five or more colors to illustrate**—a feat that took **1,000 hours of computing time to run.**

BOOLEAN LOGIC

A systematic way to write the rules of logic, this was developed by and named after the nineteenth-century mathematician George Boole. It is the basis for digital electronics and software design.

NINETEENTH CENTURY

1815 George Boole is born in Lincoln, England

1854 Boole publishes *An Investigation of the Laws of Thought*, in which he sets out the principles of Boolean logic

1864 Boole dies in Cork, Ireland

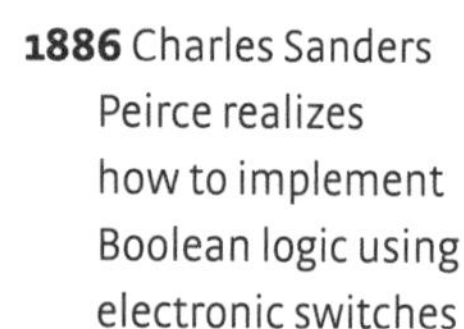

1886 Charles Sanders Peirce realizes how to implement Boolean logic using electronic switches

1965 Fuzzy logic is introduced by logician Lotfi Zadeh to handle partial truths

1995 Researchers build the first quantum logic gates, to be used in future quantum computers

WHAT IS IT?

Boole's logic concerned **"propositional logic"—whether or not statements are *true* or *false***. Statements can take the form: "All dogs are brown," which takes the value false, or "London is the capital of England," which takes the value true.

Boole **encoded "true" and "false" in variables** and then **devised ways to combine them**—in much the **same way that ordinary algebra assigns numbers to symbols, and sets rules for how the symbols can then be manipulated**2

LOGIC GATES

Two logical variables can be combined using **logic gates**. For example, an **"AND" gate** combines two Boolean variables X and Y to produce a third variable, Z, according to the following table:

Here "true" is represented by the digit 1, and "false" by 0.

Inputs		Output
X	Y	Z
0	0	0
0	1	0
1	0	0
1	1	1

Other **logical functions** exist, such as **"OR"**:

Inputs		Output
X	Y	Z
0	0	0
0	1	1
1	0	1
1	1	1

And **"NOT"**:

Inputs	Output
X	Z
0	1
1	0

THE RIEMANN HYPOTHESIS

Put forward by Bernhard Riemann in 1859, this statement about the zeroes of the Riemann zeta function could hold the key to the distribution of prime numbers. Now prove it.

THE RIEMANN ZETA FUNCTION

The Riemann hypothesis hinges around a mathematical entity called the **Riemann zeta function**. It is defined as:

$$\zeta(s) = \sum_{n=1}^{\infty} \frac{1}{n^s} = \frac{1}{1^s} + \frac{1}{2^s} + \frac{1}{3^s} + \ldots$$

In general, the **function can take complex values** (that is, with a **real** and an **imaginary component**), as can its input variable.

THE ZEROES OF $\zeta(s)$

The Riemann zeta function, $\zeta(s)$, is zero whenever s is a **negative even integer** (e.g., −2, −4, −6). These are known as the **trivial zeroes** of $\zeta(s)$.

But there are other non-trivial zeroes when *s* takes other values. The Riemann hypothesis states that for every **non-trivial** zero of $\zeta(s)$, the **real part** of s is ½.

In other words, all the non-trivial zeroes of $\zeta(s)$ occur when $s = ½ + xi$, where x is an **arbitrary real number** and i is the **unit imaginary number**.

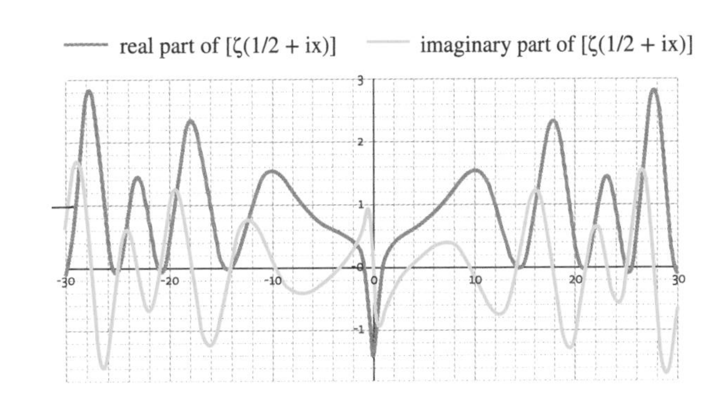

NINETEENTH CENTURY

PRIME NUMBERS

If the Riemann hypothesis is correct, then the Riemann zeta function **encapsulates the distribution of prime numbers**. In particular, the number of primes less than or equal to any number *x* is:

$$\pi(x) = R(x) - \sum_{\rho} R(x^{\rho}),$$

where

$$R(x) = \sum_{n=1}^{\infty} \frac{\mu(x)}{n} \mathrm{li}\,(x^{1/n})$$

Here, $\mu(x)$ is the **Mobius function**, li is the **logarithmic integral**, and (crucially) ρ refers to the non-trivial zeroes of $\zeta(s)$ for which the real part of $s = ½$.

x	π(x)
10	4
100	25
1000	168
10000	1229
100000	9592
1000000	78,498

For this reason, it's been called **the most important unsolved math problem**.

MILLENNIUM PRIZE

The **Clay Mathematics Institute** lists the Riemann hypothesis as one of its **seven Millennium Problems**. As such, there is a **$1 million prize** for the **first person who can provide an acceptable proof**.

FLORENCE NIGHTINGALE

Famous as "the lady with the lamp," Florence Nightingale was also a gifted mathematician and a master in the graphic representation of statistics.

THE ROSE DIAGRAM

In 1854, Florence Nightingale took a team of nurses to the Crimea to tend sick and wounded British soldiers. Over the course of two years, she kept **records of mortality rates and cause of death** among patients in the **Barrack Hospital, Scutari**. Nightingale illustrated her **data** in the form of a **pie chart**, now known as the **Rose diagram**.

Nightingale knew that vastly more men were dying from infections than from wounds, which the **chart clearly illustrated**. By implementing **good medical hygiene**, the **death toll** at the hospital was **reduced** to 2 percent. The British Army was so impressed by Nightingale's work and **statistical evidence** that it established new medical and sanitary science departments to improve healthcare.

April 1855 to March 1856

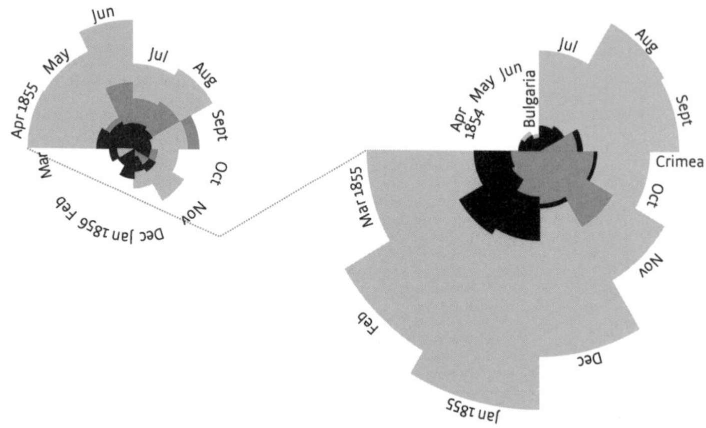

Pink wedge: *number of men who died of infections*
Red wedge: *deaths caused by wounds*
Black wedge: *deaths from all other causes*

DID YOU KNOW?

Florence Nightingale was born in 1820 in **Florence, Italy,** into a wealthy family on the **Grand Tour**.

She was educated at home by her **father William**, and enjoyed mathematics.

As a child, she made **tables recording the contents of her vegetable garden**.

She **wrote several books** on **healthcare**, **nursing**, and **hygiene**.

She **died in her sleep** on August 13, 1910, aged ninety.

Today, **data visualization** is a crucial strand of **statistics** and **data science**.

SOFIA KOVALEVSKAYA

As well as making contributions to the theory of differential equations and mathematical analysis, this Russian mathematician broke huge boundaries for women across Europe.

A EUROPEAN GENIUS

Moscow
Kovalevskaya is born on January 15, 1850. Her father works as a lieutenant general in the Imperial Russian Army. She is **educated by governesses and private mathematics tutors**, who noticed her early enjoyment of **calculus**

Heidelberg
Women are banned from Russian universities, so Kovalevskaya and her husband **Vladimir**, a paleontology student, move to **Heidelberg** in 1869, where she **gains permission to attend physics and mathematics classes at the university**

London
Kovalevskaya and her husband Vladimir visit **London**, where they move in prestigious circles. He works with the anthropologist **Thomas Huxley** and **Charles Darwin**, while Kovalevskaya meets intellectuals, including the author **George Eliot**

Berlin
On moving to Berlin in 1870, Kovalevskaya is **refused entry to Berlin University**, so she **continues her studies with Karl Weierstrass**, known as the **"father of modern analysis"**

Göttingen
Kovalevskaya presents her **doctoral thesis at the University of Göttingen** on **elliptic integrals**, the **dynamics of Saturn's rings, and partial differential equations** (the **Cauchy-Kovalevskaya theorem**). In 1874, she becomes the **first woman in modern times to be awarded a mathematics doctorate from a European university**

Stockholm
In 1883, Kovalevskaya is **offered a post as a lecturer in mathematics at Stockholm University**. Six years later, she becomes the **first woman to achieve a full professorship in mathematics at a North European university**

FACTS AND FIGURES

In **1878**, she had a **daughter Sofia** (also called "**Fufa**").

She wrote **two plays**, a **memoir** and a **novel** called ***Nihilistic Girl.***

In **1884**, she became an **editor of *Acta Mathematica*,** a scientific journal.

She **died** on **February 10, 1891,** at **forty-one** after contracting **influenza** in **Nice**.

ABSTRACT ALGEBRA

Group theory is just one example of what mathematicians call "abstract algebra"—the application of algebraic principles to general mathematical structures, such as logic and geometry.

NINETEENTH CENTURY

- **1832** Évariste Galois puts forward the first ideas in group theory
- **1847** George Boole publishes what is now known as Boolean algebra
- **1854** British mathematician Arthur Cayley comes up with the first definition of an abstract group
- **1871** German mathematician Richard Dedekind introduces the concepts of rings and fields
- **1921** German mathematician Emmy Noether gives the modern definition of a ring
- **1958** The foundations of modern algebraic geometry are set out by French mathematician Alexander Grothendieck

ALGEBRAIC STRUCTURES

When we learn **algebra in high school**, it involves simply replacing numbers with symbols. Those symbols can then be acted on by the **usual operations of arithmetic** (+, −, ×, ÷, etc.), enabling us to **solve equations**.

Abstract algebra, as the name suggests, **broadens the remit** beyond numbers **to more abstract mathematical objects**. These objects are **known, in general, as algebraic structures.**

GROUPS AND LOGIC

Algebraic structures are defined axiomatically, rather like **groups**—and groups are an example.

But there are other **structures** too; some are more complicated—others less so.

Boolean algebra, the **algebra of logic**, is an example.

Gate	Symbol	Operator
and		A.B
or		A + B
not		$\bar{A}$
nand		$\overline{A.B}$
nor		$\overline{A+B}$
xor		A ⊗ B

RINGS AND FIELDS

Two other algebraic structures, for which abstract algebras may be composed, are known as "rings" and "fields."

Rings are an extension of groups that permit addition, subtraction, and multiplication. The **integers** are an example of a ring.

Fields admit addition, subtraction, and multiplication, and also division. The inclusion of division **means that the real numbers** (including but not limited to the integers) **are an example of a field**.

Every possible kind of geometry gives rise to its own **"polynomial ring"**—these algebraic structures form the basis for an **algebraic approach to geometry**.

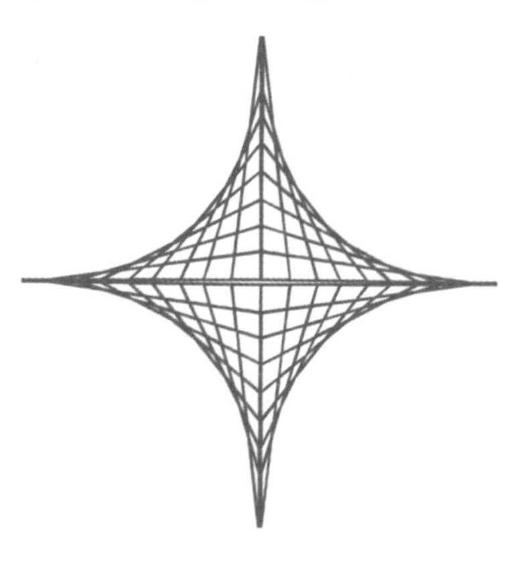

GEORG CANTOR

Georg Cantor was a troubled genius who suffered severe bouts of depression. He pioneered set theory, within which infinity becomes a real concept rather than a hypothetical one.

March 3, 1845 Cantor is born in St. Petersburg, Russia. As a child, he is an exceptional violinist

1856 After suffering illness, Cantor's father moves the family to Frankfurt, where the winters were milder. Young Cantor excels in mathematics, trigonometry in particular

1862 He studies at the Swiss Federal Polytechnic in Zurich and the universities of Berlin and Göttingen, where he is awarded his doctorate in 1867

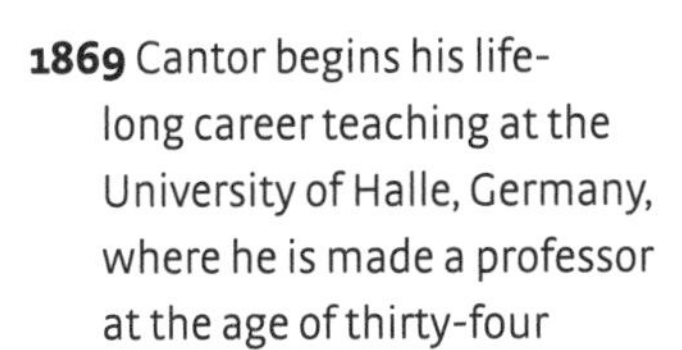

1869 Cantor begins his life-long career teaching at the University of Halle, Germany, where he is made a professor at the age of thirty-four

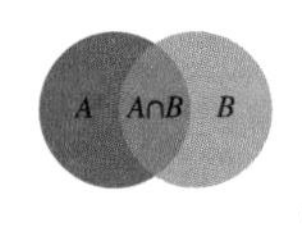

1874 Cantor begins his life's work on set theory, the mathematics describing collections of objects. It establishes a whole new realm—the mathematics of the infinite

1874 Cantor marries Vally Guttmann and they have six children

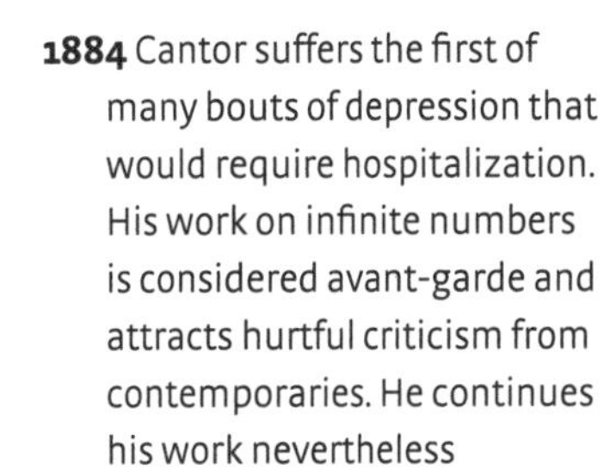

1884 Cantor suffers the first of many bouts of depression that would require hospitalization. His work on infinite numbers is considered avant-garde and attracts hurtful criticism from contemporaries. He continues his work nevertheless

1890 Cantor becomes the first president of the German Mathematical Society

1899 His son Rudolph dies at age twelve, which plunges Georg into another bout of depression

January 6, 1918 Cantor dies of a heart attack in a sanatorium

A BITTER CONTROVERSY

Cantor's theories on infinite numbers horrified his peers, in particular his former professor **Leopold Kronecker**, who held that "God made the integers, all the rest is the work of man"—meaning **infinity is just a construct**. The **criticism weighed heavily on Cantor** and **precipitated his depression**.

SET THEORY

Concerned with the organization and classification of objects, set theory is a foundational pillar of mathematics and has applications in database programming and the theory of probability.

NINETEENTH CENTURY

1874 Georg Cantor initiates the development of set theory with his work on infinite sets

1880 John Venn introduces his visual representation of sets

1901 British philosopher Bertrand Russell points out a paradox of set theory in the form of a contradiction, the "set of objects not in a set"

1908 German logician Ernst Zermelo proposes the first attempt to develop a new paradox-free set theory

1922 Zermelo's work is supplemented and expanded to become a working theory by Abraham Fraenkel and Thoralf Skolem

SET NOTATION

Sets are **used to label groups of objects**—cars, mathematical functions, ancient languages, anything. Set theory **uses the following notation:**

$\in$ means **"is a member of."** So if x denotes a person and the set of all people is denoted A, then we can say $x \in A$.

$\subseteq$ means **"is a subset of."** If B denotes the set of people who live in France, then $B \subseteq A$.

$\cap$ means **"intersection between,"** or the common ground between two sets. If C is the set of everything in France, then B is just $A \cap C$.

$\cup$ means **"union of,"** or the merger of two sets. So if D is the set of people not in France, then A is given by $B \cup D$.

VENN DIAGRAMS

Sets can be **presented graphically**, using a **diagrammatic scheme** invented in the late nineteenth century by British mathematician **John Venn**.

Sets are represented by **regions of the diagram**, which can **overlap to form unions and intersections**.

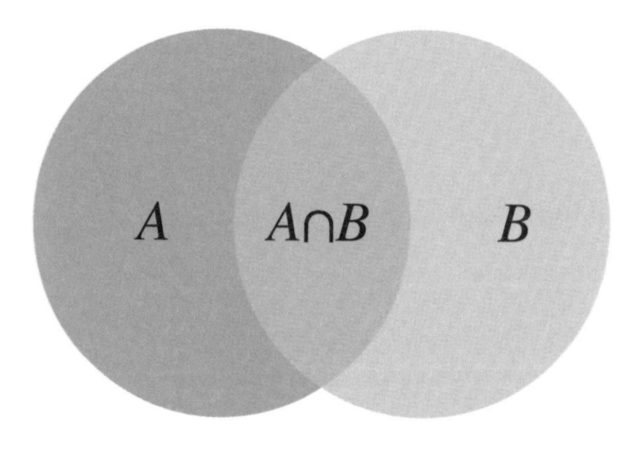

HENRI POINCARÉ

A brilliant polymath, Poincaré made seminal contributions to the fields of topology, celestial mechanics, relativity, differential equations, and group theory.

CHILDHOOD AND EDUCATION

Poincaré was born into a high-flying family in 1854 in **Nancy**, France. His **father was a professor of medicine at the University of Nancy** and his **cousin Raymond Poincaré would serve as President of France** from 1913 to 1920.

Young Poincaré became **seriously ill with diphtheria during childhood**, but he recovered and attended the **Lycée Nancy (secondary school)**. He studied mathematics at the **École Polytechnique**, Paris.

In 1879, Poincaré was **awarded his doctorate in the field of differential equations**. He worked as a **mining and railway engineer** while also **teaching at the University of Caen and at the Sorbonne**.

He garnered **many awards** in his career, including the **Gold Medal of the Royal Astronomical Society of London (1900)**.

Poincaré **died in 1912 from an embolism** after prostate surgery. He was fifty-eight.

GREATEST WORK

Poincaré established a **new branch of mathematics**—the **qualitative theory of differential equations**. This examines the **properties of differential equations** by **looking at factors other than their solutions**.

Poincaré studied the **motion of the electron** and formulated a theory that **almost anticipated Einstein's discovery of special relativity** for fast-moving bodies.

He analyzed the **three-body problem** in gravitational physics, with a view to answering the question of **whether the Solar System is stable**. In doing so, he showed that the **motion of planets could be chaotic**.

Poincaré was fascinated by **topology** and advanced it by using **algebraic formulations**.

He **wrote articles and books on physics and mathematics for the general reader**, aiming to **popularize science** and boost the public understanding of it.

HERMANN MINKOWSKI

One of Einstein's tutors, Minkowski is famous for having shown that his former student's special theory of relativity could be interpreted geometrically as a 4-D space-time.

RUSSIAN GENIUS

Minkowski was born on June 22, 1864, in **Aleksotas**, in the Russian Empire (now part of modern **Lithuania**). **His parents were both Jewish**.

To escape **pogroms** in Russia, the family moved to the Prussian city of Königsberg in 1872. A **child prodigy**, Minkowski went up to university at the age of **fifteen**, studying at **Königsberg** and **Berlin**.

While still a student, Minkowski **won the prestigious Grand Prix des Sciences Mathématics** of the **French Academy of Sciences**.

$E = mc^2$

Minkowski **taught at the universities of Bonn, Göttingen, Königsberg**, and at the **Eidgenössische Polytechnikum**, where Einstein was one of his students.

In 1896, he published his ***magnum opus* in number theory, *Geometrie der Zahlen* (*Geometry of Numbers*)**.

He married **Auguste Adler** in 1897 and they had **two daughters**.

On January 12, 1909, Minkowski **died suddenly of appendicitis**.

SPECIAL RELATIVITY

In 1905, **Einstein** presented his **special theory of relativity**, establishing the **relationship between space and time** and the **effect it has on the dynamics of moving bodies**.

Minkowski realized that **Einstein's algebraic equations could be expressed and best understood in geometric terms**. In 1907, he **combined three-dimensional Euclidean space and time into a four-dimensional manifold**. His representation became known as **Minkowski space-time**.

In 1916, Einstein published his towering achievement, the **general theory of relativity**, in which the **force of gravity arises from introducing curvature to Minkowski's flat space-time**.

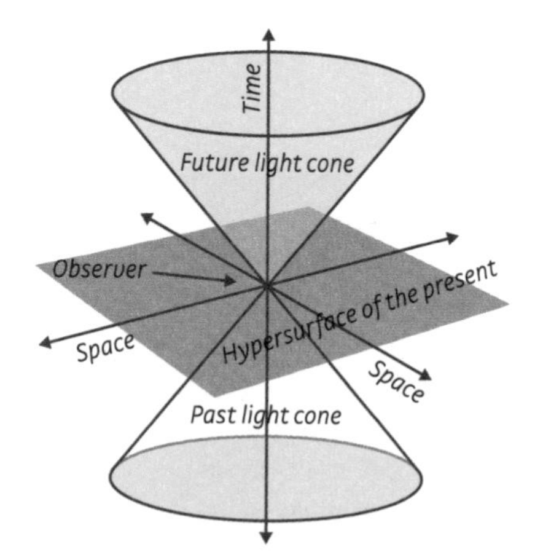

DID YOU KNOW?

Minkowski called his student Einstein a **"lazy dog"** because he thought he wasn't interested in mathematics.

TENSORS

Tensors are multidimensional arrays of numbers, rather like vectors and matrices. They are used in physics and mathematics to capture complex relationships between quantities.

- **1846** The word tensor is first coined by Irish mathematician William Rowan Hamilton
- **1890** The methods of tensor calculus are developed by the Italian mathematician Gregorio Ricci-Curbastro
- **1898** German physicist Woldemar Voigt gives the word tensor its present-day meaning
- **1915** A tensor treatment of differential geometry forms the backbone for Einstein's general theory of relativity

STRESS TENSOR

A common example in physics is the stress tensor, **used to characterize the deforming stresses acting on an object or fluid**. Indeed, the word "tensor" derives from the Latin "*tendere*," which means **"to stretch."**

The stress **tensor is a 2-D tensor** which, when **multiplied by a unit-length vector**, gives the **stress** (or force per unit area) **acting through a plane perpendicular to the vector**.

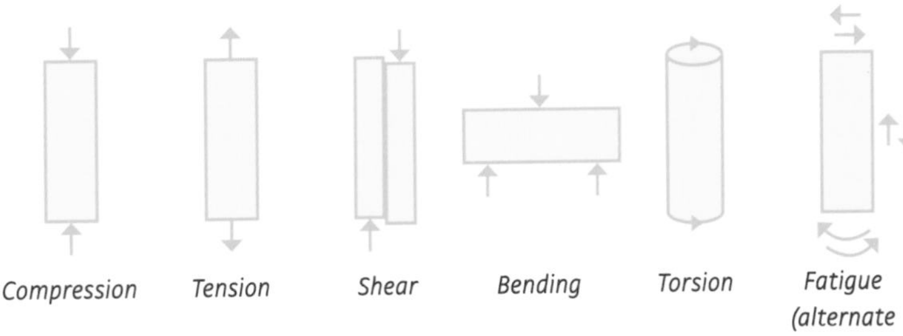

This gives the flexibility to capture the **complex stretching and squeezing forces** within a building or perhaps a current of flowing liquid.

EXAMPLE

In two dimensions, the stress tensor resembles a 2 × 2 matrix:

$$T = \begin{pmatrix} \sigma_{xx} & \sigma_{xy} \\ \sigma_{yx} & \sigma_{yy} \end{pmatrix}$$

Given a unit-length vector

$$n = \begin{pmatrix} n_x \\ n_y \end{pmatrix}$$

it acts to create a stress of $n_x \sigma_{xx} + n_y \sigma_{xy}$ in the x direction and $n_x\sigma_{yx} + n_y\sigma_{yy}$ in the y direction.

OTHER APPLICATIONS

ELECTROMAGNETISM

The Faraday tensor is constructed from the **electric and magnetic field strengths** in **Maxwell's electromagnetic theory**.

GENERAL RELATIVITY

The **Einstein tensor** encapsulates the **complex deformations to space and time** caused by the **gravitational field**.

COMPUTER VISION

Tensor-based techniques are used for **processing and manipulation of images**.

POINCARÉ CONJECTURE

One of the first theorems in topology, the Poincaré conjecture says that 3-D spherical spaces are "simply connected." It was finally proven in 2003—100 years after first being stated.

1904 The conjecture is first posed by French mathematician Henri Poincaré

1982 American mathematician Richard Hamilton develops the geometrical technique of Ricci flow

1986 American Michael Freedman wins the Fields Medal for proving the conjecture for a four-sphere

2003 A proof of the Poincaré conjecture for a three-sphere, using Ricci flow, is published by the Russian mathematician Grigori Perelman

2006 Perelman's proof is independently confirmed

THREE-SPHERES

The spheres that we're used to in our everyday experience are **two-spheres**—that is, **2-D surfaces wrapped into the shape of a sphere**.

A **three-sphere** is **one dimension higher**.

SIMPLY CONNECTED?

To say a surface is simply connected means that **an arbitrary loop can be shrunk smoothly to a point**. Generally speaking, this is possible if there are **no holes in the surface**.

For example, an elastic band wrapped around a two-sphere can always be shifted so that it shrinks smoothly down to a point without you needing to cut the band or the sphere.

But try that on a **torus** (a shape like a ring doughnut) and it's **not always possible—due to the hole through the middle**.

POINCARÉ'S QUANDARY

It was well known that a two-sphere is simply connected. But the question remained whether that was also true of a three-sphere. Asserting that it probably was, in 1904, was Poincaré's conjecture.

PROOF

Russian mathematician **Grigori Perelman** proved that the three-sphere is indeed simply connected in 2003. He used **"Ricci flow,"** a **technique for deforming topological spaces**, developed by the American **Richard Hamilton**.

To date, the Poincaré conjecture is the only one of the **Clay Mathematics Institute's Millennium Problems** to be solved—although **Perelman declined the $1 million prize**.

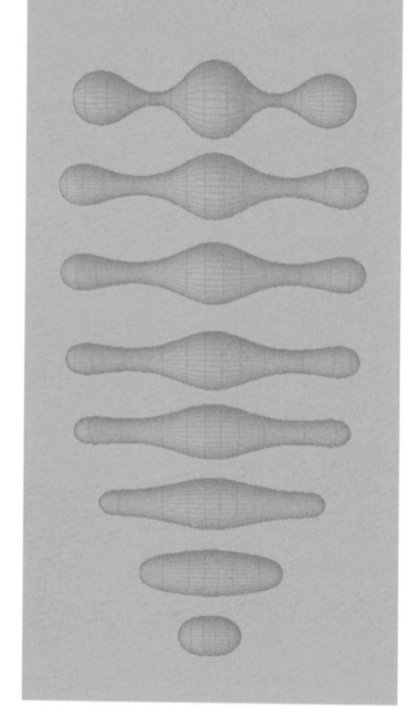

DISCRETE GEOMETRY

Constructing shapes in discrete geometry is rather like looping string around a pin board. Despite its apparent simplicity, it has found applications in software design and quantum theory.

WHAT IS IT?

Ordinary **geometry**—the mathematical study of lines, angles, shapes, and solids—is **"continuous,"** in the sense that all points in space are allowed.

In **discrete geometry**, however, **that all changes**. Here, all you get is a **well-defined lattice of points, and all other regions are off limits**.

PICK'S THEOREM

One remarkable result in discrete geometry was proved in 1899 by the Austrian mathematician **George Pick**.

He found a formula for **calculating the area of an arbitrary-shaped polygon** formed by joining points on a discrete grid, each separated by one unit.

If the number of points marking the outside of the shape is *x*, and the number of points inside the shape (not touching the boundary) is *y*, then the area inside is given by the formula:

$$A = \frac{x}{2} + y - 1$$

It's very simple and **works for any shape you care to draw**. Try it!

APPLICATIONS

POLYGONS

Discrete geometry has applications in the **pure study of geometric shapes**, such as **tessellations** and **aperiodic tilings**.

LATTICE GAUGE THEORIES

Some toy models in **quantum physics** working by **discretizing space and time** into a **grid of points a fixed distance apart**.

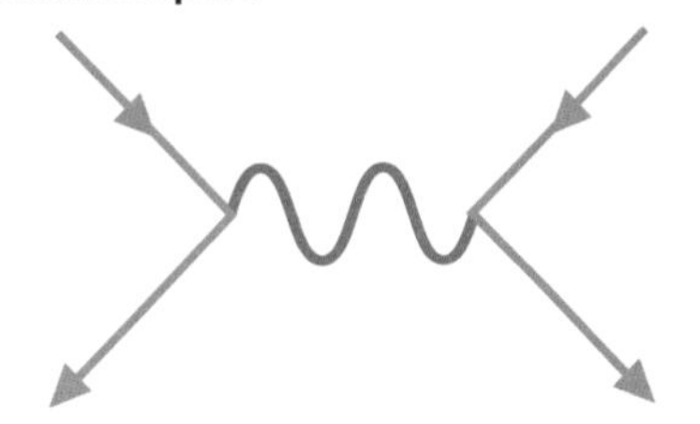

PACKING PROBLEMS

Attempts to calculate the **most efficient way to stack shapes and objects together**, such as **Kepler's conjecture**, are problems made for discrete geometry.

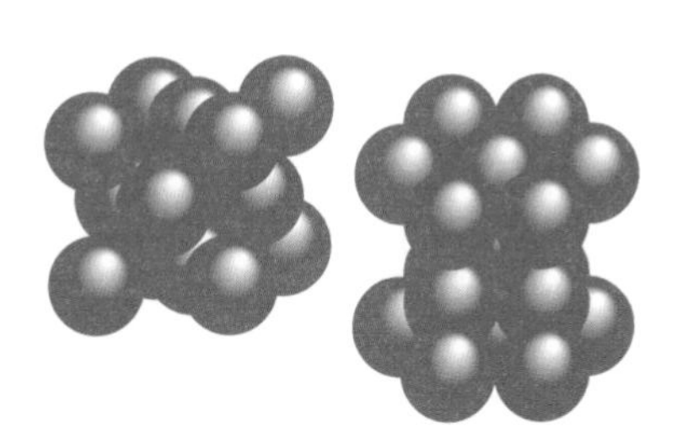

COMPUTER GRAPHICS

Less obvious in the high-definition world of today, but computer graphics are also an **elaborate construct of discrete geometry**.

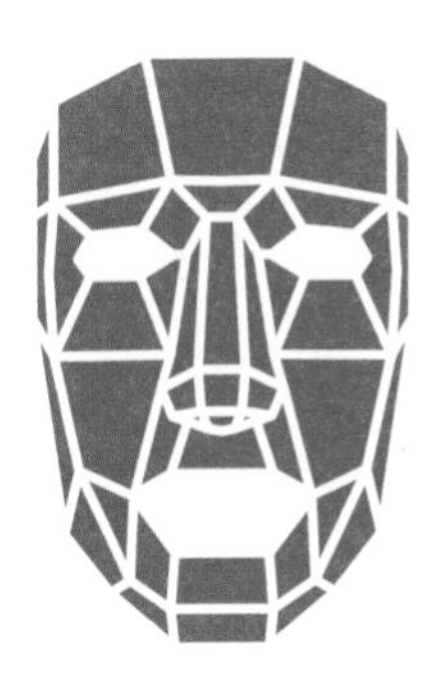

DAVID HILBERT

Part of a mathematical golden age in Germany, Hilbert established axioms and methods that are still highly influential today.

January 23, 1862 David Hilbert is born in Königsberg, Prussia, on the Baltic Sea. The city is now called Kaliningrad and is part of Russia

1860s His mother, Maria, a keen amateur mathematician and astronomer, encourages her son to study these pursuits

1880 Hilbert studies at the University of Königsberg, where he later lectures. He befriends Hermann Minkowski, a groundbreaking mathematician, in what proves to be a fertile collaboration

1896 Hilbert becomes a professor at the prestigious University of Göttingen in Germany. Famous alumni and professors include Carl Friedrich Gauss, Peter Dirichlet, and Bernhard Riemann

1930 After suffering from pernicious anemia, Hilbert retires from teaching. Sadly, many of his Jewish colleagues are forced to flee Göttingen during the Nazi purges

1943 Hilbert dies on February 14. He was a friendly, passionate, and optimistic man, who inspired gifted students. His epitaph reads "We must know. We will know"

KEY ACHIEVEMENTS

In 1888, Hilbert made a huge advance by proving his **basis theorem**, an idea from **abstract algebra** concerning the **properties of algebraic structures** known as **rings**.

Hilbert formulated a **definitive set of axioms** for **Euclidean geometry**, which he published to acclaim in 1899.

He put forward a **set of problems** in **geometry**, **algebra**, and **calculus**, known as **Hilbert's twenty-three problems**. He believed that if these could be solved, mathematics could be greatly advanced. The list included the **Riemann hypothesis**, which concerns the **structure of the Riemann zeta function** and remains **unsolved to this day**.

$$\zeta(s) = \sum_{n=1}^{\infty} \frac{1}{n^s}$$

Hilbert was **one of the founders of mathematical logic**. His work on techniques to prove mathematical truths **established rigorous working methods** that are **still in use today**.

THE PARETO PRINCIPLE

In a nutshell, the Pareto principle basically states that approximately 80 percent of results come from 20 percent of causes.

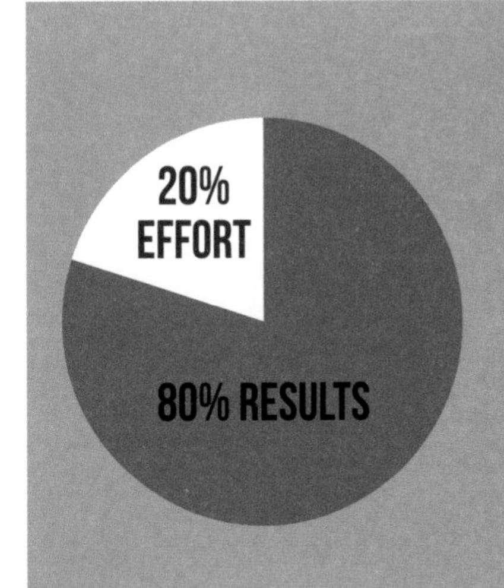

PEA PLANT ORIGINS

The principle was named after the Italian economist **Vilfredo Pareto** (1848–1923) who noticed that 20 percent of the **pea plants** in his garden generated 80 percent of healthy pea pods.

He began to apply this rule broadly across society and conducted studies showing that around 80 percent of the **land in Italy** was owned by 20 percent of the populace. Pareto carried out observations in **other countries** and found a **similar distribution**.

Pareto had discovered a recurring pattern that crops up all around us; for example, in terms of **population density**, **business efficiency**, and **consumer spending habits**. It's also known as the **eighty–twenty rule**.

EXAMPLES OF THE EIGHTY–TWENTY RULE

Twenty percent of sales reps generate 80 percent of sales.

Twenty percent of software bugs cause 80 percent of crashes.

Twenty percent of customers account for 80 percent of profits.

Eighty percent of tax is paid by 20 percent of the population.

PARETO DISTRIBUTION

In **statistics**, the Pareto distribution is a **probability distribution** with a **power-law form**,

$$P(x) = \frac{\alpha x_m^\alpha}{x^{\alpha+1}}$$

where $\alpha > 0$, and x_m is the lower bound of x. This produces a distribution with a slowly declining tail.

The Pareto principle corresponds to the special case when $\alpha = \ln(5) / \ln(4) \approx 1.161$.
Where ln is the logarithm to base e.

Sociological data often conforms to the Pareto distribution. For example, the **largest percentage of people live in a small number of cities**.

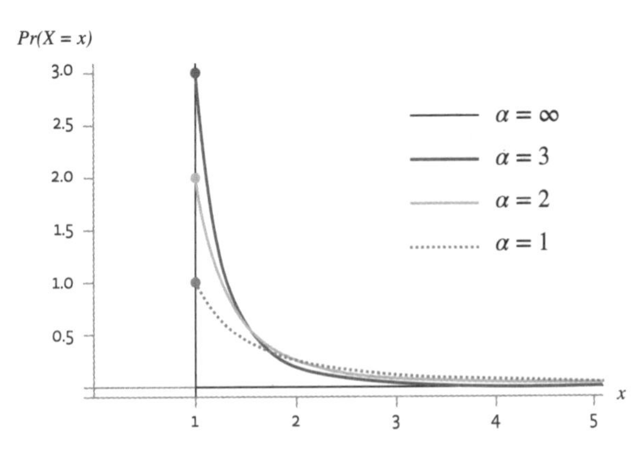

MARKOV PROCESSES

Markov processes, sometimes known as "random walks," occur when a mathematical variable evolves randomly. They crop up in physics, gambling, and the study of financial markets.

RANDOM WALK

A **bee buzzing from plant to plant or a drunk meandering between lamp posts**, trying to find his way home, are all examples of a *random walk*.

For example, start at a point in the middle of a 2-D grid and at each step move randomly in one of the four possible directions (up, down, left, or right).

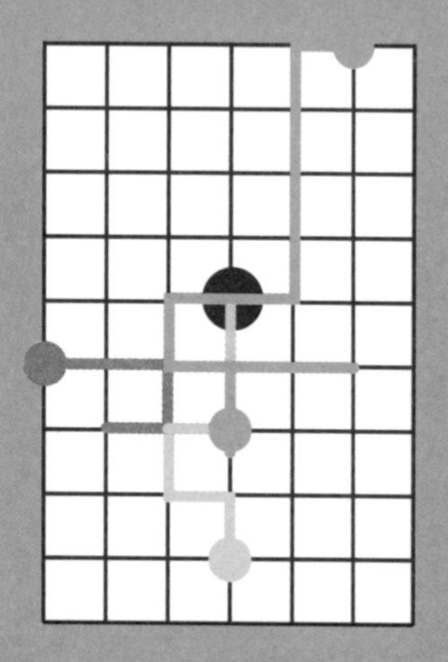

On average, after n steps, the walker will have wandered a distance of $\sqrt{n}$ steps from their starting point.

MEMORYLESSNESS

When the probabilities that a random walker will move in any given direction don't change with time, they are said to be *memoryless*. This is also known as the **Markov property**, after the Russian mathematician **Andrey Markov**, who discovered it in 1906.

A memoryless random walk is known as a **Markov process**, or **Markov chain**.

EXAMPLE

Imagine a process that randomly switches between two states, labeled A and B.

Let's say, if the process is in state A then, at the next step, it stays where it is with a **probability** of 0.6 or moves to state B with a probability of 0.4. Similarly, if it's in state B, it stays where it is with probability 0.3 or moves to A with probability 0.7.

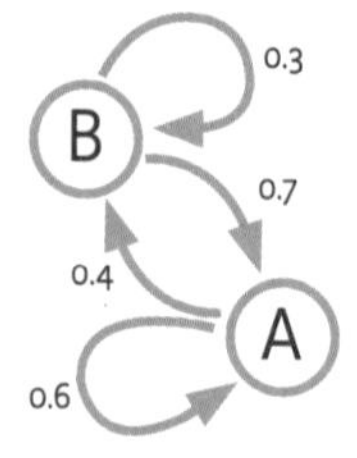

Since these four probabilities do not change as the state evolves, this can be considered a **Markov process**.

APPLICATIONS

Gambling: if you repeatedly **bet** on the outcome of a **coin toss**, your bank balance will undergo a **Markov process**.

Physical sciences: Markov chains are used to model the **evolution of processes in quantum physics** and **chemistry**.

Statistical modeling: Markov processes describe many problems in **applied statistics** and **data analysis**.

SRINIVASA RAMANUJAN

In his short life, Srinivasa Ramanujan made contributions to analysis, number theory, and infinite series. With G.H. Hardy, he discovered the first taxicab number: 1729.

December 22, 1887 Ramanujan born in Erode, Madras, India. In childhood, he survives smallpox but later suffers bouts of dysentery

1892–1907 Ramanujan excels in mathematics at school and college. He has little interest in any other subject and twice fails his Fellow of Arts exam

1910 Becomes a researcher at the University of Madras

1913 Ramanujan writes to G.H. Hardy, a Cambridge lecturer, presenting some of his theorems. Hardy recognizes genius and invites him to study at Trinity College, under a scholarship. Ramanujan goes on to spend five years at Cambridge

1918 Becomes the first Indian person elected as a fellow of Trinity College, Cambridge. He also becomes a member of the Royal Society

April 26, 1920 After contracting tuberculosis, Ramanujan returns to Madras, where he dies aged thirty-two

TAXICAB NUMBERS

G.H. Hardy visited Ramanujan in a London hospital and traveled in a taxi with the number 1729. Hardy remarked it was "rather a dull number," but Ramanujan contradicted him, saying "No, it is a very interesting number. It is the smallest number expressible as the sum of two cubes in two different ways."

The two ways are:
$1729 = 1^3 + 12^3 = 9^3 + 10^3$

1729 is also known as the "Hardy–Ramanujan number."
In general, the taxicab numbers, Ta(n), are the smallest numbers that can be expressed as the sum of two cubes in n distinct ways.

n	Ta(n)	sums of cubes
1	2	$1^3 + 1^3$
2	1729	$1^3 + 12^3$
		$9^3 + 10^3$
3	87539319	$167^3 + 436^3$
		$228^3 + 423^3$
		$255^3 + 414^3$
4	6963472309248	$2421^3 + 19083^3$
		$5436^3 + 18948^3$
		$10200^3 + 18072^3$
		$13322^3 + 16630^3$
5	48988659276962496	$38787^3 + 365757^3$
		$107839^3 + 362753^3$
		$205292^3 + 342952^3$
		$221424^3 + 336588^3$
		$231518^3 + 331954^3$
6	24153319581254312065344	$582162^3 + 28906206^3$
		$3064173^3 + 28894803^3$
		$8519281^3 + 28657487^3$
		$16218068^3 + 27093208^3$
		$17492496^3 + 26590452^3$
		$18289922^3 + 26224366^3$

G.H. HARDY

G.H. Hardy championed the study of pure mathematics, in particular analysis and number theory. His contributions touched physics, genetics, and the public appreciation of mathematics.

Godfrey Harold Hardy was born in 1877 in **Cranleigh, Surrey**. His mathematical talent shone from an early age. **As a child**, he would amuse himself by **factorizing the numbers of hymns** in church.

In 1899, Hardy graduated with a B.Sc. in mathematics from **Trinity College, Cambridge**, where he went on to become a **fellow and lecturer**. In 1931, he was appointed **Sadleirian Professor of Pure Mathematics at Cambridge**, where he remained until his death in 1947. He was a **lifelong pacifist** and **abhorred the use of mathematics in military applications**.

CAREER HIGHLIGHTS

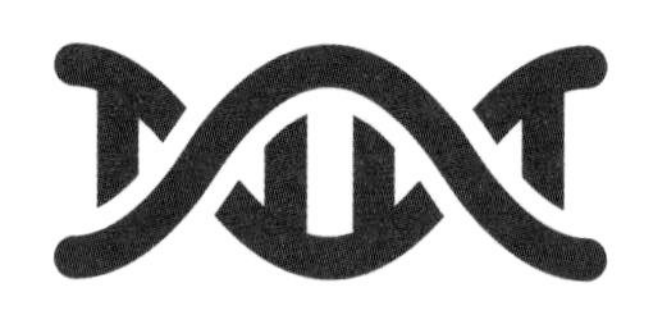

In 1908, Hardy and the German physician **Wilhelm Weinberg** demonstrated a law in population genetics, which later became known as the **Hardy–Weinberg principle**. It states that the **fraction of alleles** (variants of specific genes) **remains constant from generation to generation if there are no other evolutionary influences**. This **debunked the theory that dominant alleles would always become more common**.

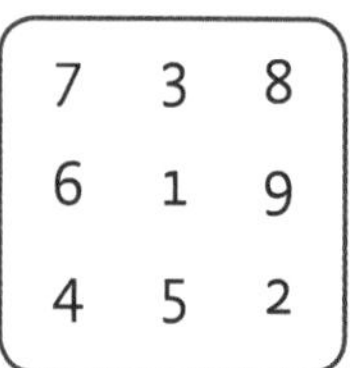

Hardy collaborated extensively with fellow Cambridge don **John Edensor Littlewood** to achieve **major breakthroughs in number theory**. This led to a **series of eponymous conjectures**.

Hardy mentored **Srinivasa Ramanujan**, the famous Indian mathematician. Their work on **integer partitions** has been applied in **physics**.

He wrote **300 papers** and **eleven books**, including ***A Course of Pure Mathematics*** (1908), which became a required university text.

Hardy wrote the popular essay ***A Mathematician's Apology*** in 1940, which offered insight into the mind of a pure mathematician. He defended the value of **pure mathematics** and explained its beauty, **akin to poetry and painting**.

In 1947, he was awarded the **Royal Society's** prestigious **Copley Medal** for the development of **mathematical analysis**.

EMMY NOETHER

This brilliant scholar overcame sexism to work in the field of abstract algebra and proved Noether's theorem, linking conserved quantities to symmetries in physics.

WHO WAS EMMY NOETHER?

On March 23, 1882 Amalie Emmy Noether was born in **Erlangen, Bavaria**, Germany. Her **father, Max**, was a mathematician and the family was **Jewish**. In 1900, Noether became **one of only two female students** at the **University of Erlangen**.

She also studied at the **University of Göttingen**, a world-leading center for mathematics, where lecturers **David Hilbert** and **Felix Klein** recognized her genius. In 1907, she gained her **Ph.D. in algebraic invariants**, an area of **group theory**.

Despite opposition from some academics, Noether **lectured** for many years at **Erlangen's mathematical institute and at Göttingen**. She worked **unpaid for long periods in her career**.

In 1933, she was forced to flee Germany due to **Nazi persecution** and was offered a position at **Bryn Mawr Women's College in Pennsylvania**.

She **died on April 14, 1935**, following an operation to remove a tumor. Tributes flooded in from scientists across the globe.

ABSTRACT ALGEBRA

Noether was fascinated by abstract algebra, which is the study of **algebraic structures**, such as **groups**, **rings**, **modules**, **vector spaces**, and **lattices**. Her work in this field was **highly influential**.

NOETHER'S THEOREM

Symmetry	Conservation law
Translation in space	Linear momentum
Translation in time	Energy
Rotation in space	Angular momentum
Gauge transformations	Electric, weak & color charge

In 1915, Noether was called upon by Hilbert and Klein to solve a problem with **Einstein's theory of general relativity. It didn't adhere to the established principle that energy can change form but never be destroyed**.

Noether proved that **energy** was **conserved**, just in a slightly different way. She proved that energy conservation followed from time symmetry and linked other conserved quantities to symmetries in nature.

FRACTALS

Fractals are disjointed shapes with a captivating beauty. The geometry of nature itself, they are generated by rigid mathematical laws and have some astonishing properties.

WHAT ARE FRACTALS?

At school, we learn about **Euclidean geometry and the construction of shapes with "smooth" outer lines**. Fractals, however, are **"rough."**

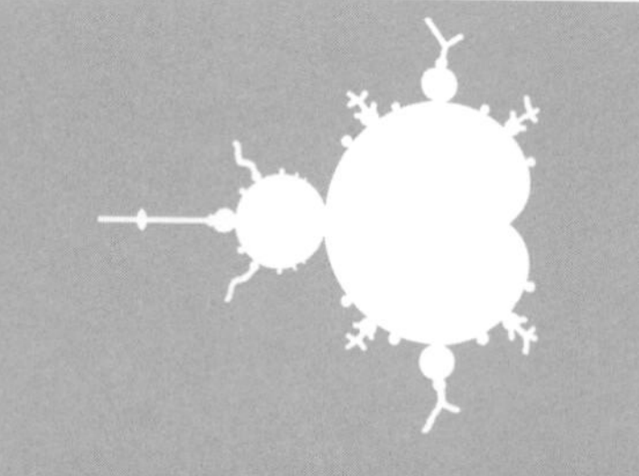

They have the property of **"self-similarity,"** in that **zooming in** simply **reveals repeated copies of the original fractal**.

The term fractal was **first used** in 1975 by the Belgian mathematician **Benoit Mandelbrot**, from the Latin word ***fractus***, meaning **"broken."**

THE KOCH SNOWFLAKE

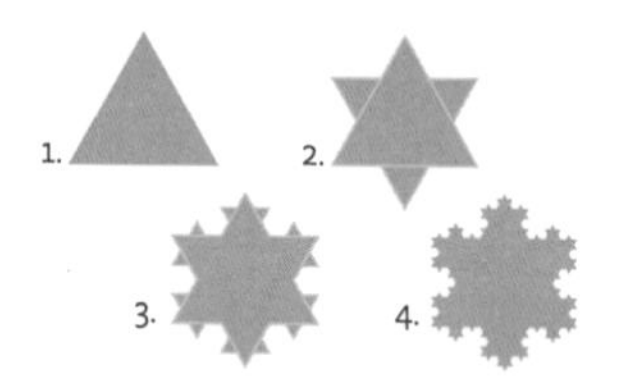

In 1904, the Swedish mathematician **Helge von Koch** described one of the earliest known examples of a fractal.

- Start with an **equilateral triangle**.
- Glue **another equilateral triangle**, one-third the size, to the middle of each side.
- **Repeat** the process **indefinitely**.

The repeated application, or **iteration**, of a rule is a **defining characteristic of fractal geometry**. And this is how fractals achieve their rough texture.

FRACTAL DIMENSION

Even though the outer edge of a **snowflake** is a **1-D line**, its **tangled structure** in some sense gives it a dimension greater than one.

The **dimension of any fractal** is defined as

$$D = -\frac{\ln N}{\ln \varepsilon}$$

where N is the number of **self-similar pieces** and ε is the **magnification**. For the Koch snowflake, each iteration leads to four self-similar pieces, at magnification 1/3. So

$$D = \ln 4 / \ln 3 \approx 1.262.$$

NATURAL SCIENCE

Fractals are used today in **digital imaging**, **computer game design**, **medicine**, **geology**, and **seismology**. Look for them in **nature**, for instance fern leaves, trees, river deltas and peacocks' tails. Fractals are also a geometrical manifestation of **chaos theory**.

GASTON MAURICE JULIA

This Algerian-born French mathematician lived between 1893 and 1978. He presented an award-winning paper in 1918 that led to an **early type** of fractal known as the **"Julia set."**

The advance of **computer technology** in the 1970s sparked renewed interest in fractals, with sets being created in **beautiful color palettes** and **mesmerizing shapes**.

RONALD FISHER

Considered to be one of the greatest scientists of the twentieth century, Fisher invented modern statistics and developed revolutionary techniques for applying it to his other love, genetics.

WHO WAS RONALD FISHER?

Born in London on February 17, 1890, Ronald Aylmer Fisher was the son of a wealthy auctioneer and fine art dealer. He studied at **Gonville & Caius College, Cambridge**, where he developed an interest in **evolutionary theory**.

In 1919, he began working as a **statistician**, observing **trials on crops** and **gathering data**.

He applied his knowledge in the statistical field to **evolution**, with ground-breaking results, and **pioneered much of modern statistics**.

Fisher **died** in Adelaide, South Australia, on **July 29, 1962**.

KEY FINDINGS

In 1918, Fisher published a key paper outlining statistical tools that could be used to **iron out anomalies** between **Darwin's theory of natural selection and the theory of genetic inheritance**.

Fisher introduced the principle of **"randomization"** to the **design of experiments**. This was embodied in the **"lady tasting tea"** experiment, where Fisher proposed to deduce whether **Muriel Bristol** really could tell **whether or not the milk had been added to the cup first** (as she claimed she could) by serving her a series of cups **randomly prepared** each way.

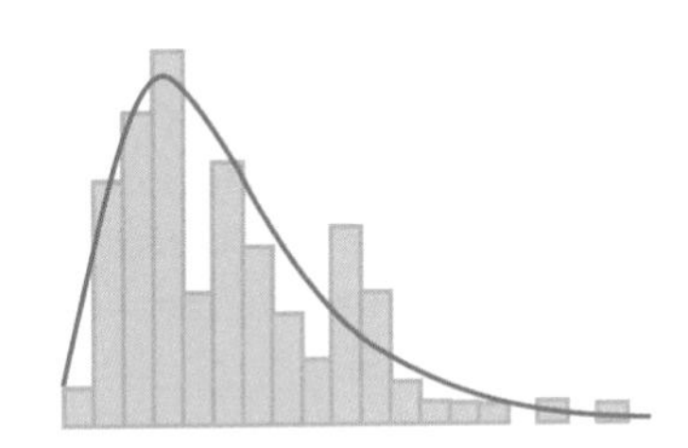

He developed the concept of **analysis of variance** (ANOVA). This is a technique for determining **whether data points in a random sample all come from the same underlying population, or a number of different ones**.

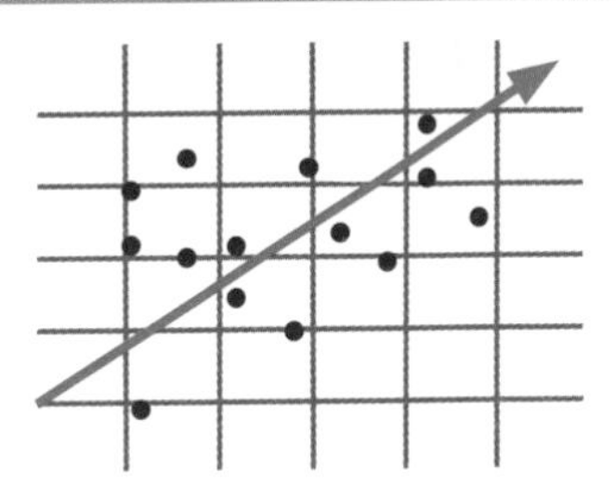

He also pioneered techniques in statistics such as **maximum likelihood for estimating parameters in statistical models**, **probability distributions**, and **hypothesis testing** for assessing the truth (or not) of statements given data.

In 1925, he published the book ***Statistical Methods for Research Workers***, which became a go-to text for **scientists in all fields** for fifty years.

P-VALUES

A p-value is a statistic calculated from experimental data to test whether a hypothesis might be true or false. Yet many statisticians believe p-values are misused and misinterpreted.

WHAT ARE THEY?

Usually they are **framed in terms of a hypothesis**, *H*, which proposes some kind of **connection between quantities** being studied.

The **p-value** is the **probability that we would see experimental results equal to or more extreme than those observed if there were no such connection**. The theory that there is no connection between the observed quantities is sometimes called the **null hypothesis,** H_o.

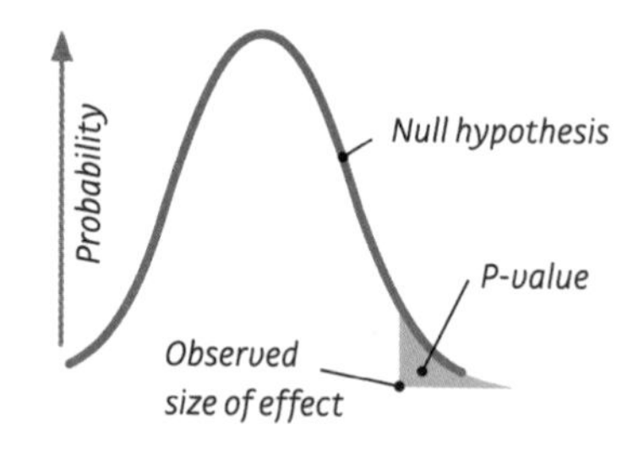

In other words, the **p-value is the probability of the results being a fluke**.

STATISTICAL SIGNIFICANCE

In the past, p-values have been used as the acid test for the **statistical interpretation of experimental data**.

In particular, a p-value of 0.05 or less—the **probability of seeing results this extreme by pure chance** being less than or equal to 5 percent—has become the **de facto criterion** to claim that the evidence for *H* is **statistically significant**.

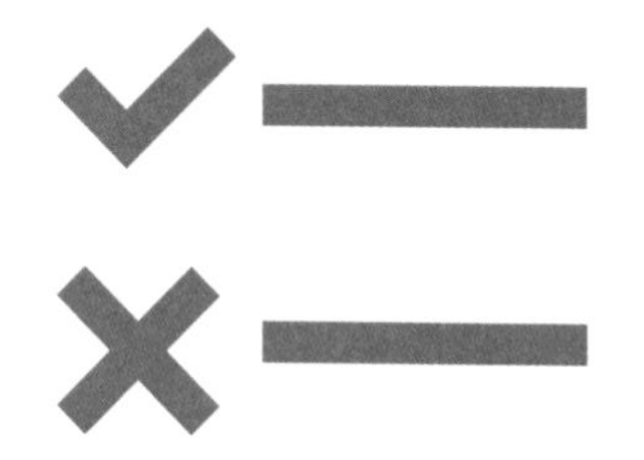

WHAT'S WRONG WITH THEM?

The problem is two-fold.

First, the **0.05 threshold for significance is arbitrary**. If you come up with twenty alternative hypotheses, then, **on average**, one of them will have a p-value that appears statistically significant.

And second, the **p-value is not the probability that *H* is false**. So a **low p-value doesn't necessarily mean a high probability that your hypothesis is true**.

Truth table	H_o true	H_o false
Test says Accept H_o	☺	☹
Test says reject H_o	☹	☺

REPLICATION CRISIS

Misuse and poor interpretation of p-values is thought to be a contributing factor in the so-called **replication crisis**—the fact that **attempts to replicate many published studies in psychology and social sciences have failed**. The implication is that the **statistical analysis**, used to infer results in the original studies, was **flawed**.

MODERN STATISTICAL INFERENCE

P-values are widely regarded as a flawed statistical framework for interpreting experimental data. An alternative technique is based on Bayes's theorem from probability theory.

BEYOND P-VALUES

Part of the problem with p-values is that they tell you the **conditional probability** of seeing the data, *D*, given that the hypothesis, *H*, is true, i.e., $p(D|H)$. But **what scientists really want to know** is $p(H|D)$—the probability that the hypothesis is true given the data.

These two quantities are related by Bayes's theorem.

$$p(H|D) = \frac{p(D|H)p(H)}{p(D)}$$

And this enables calculation of $p(H|D)$ (called the ***posterior* probability**) in terms of $p(D|H)$ (called the **likelihood**, or **evidence**) and our previous expectations about the hypothesis (called the ***prior* probability**).

BAYESIAN INFERENCE

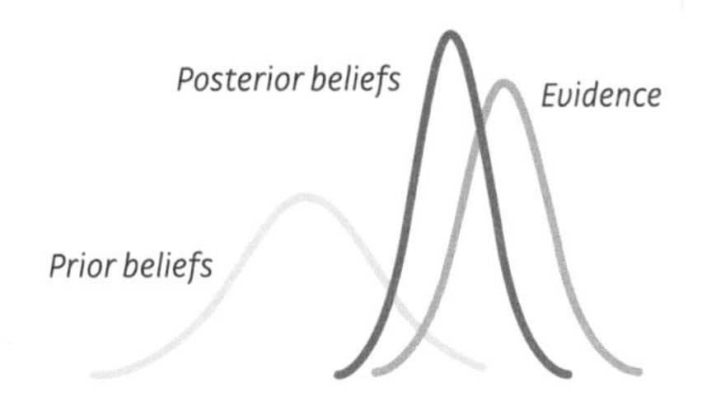

Let's say the hypothesis boils down to the **value of a parameter** in a model. Then the **prior beliefs** represent a **probability distribution** for what we initially believe about the value of the parameter, with the **peak of the distribution** giving the **most likely value**. The **evidence then gives us another distribution**. And Bayes's theorem tells us how we should **aggregate the prior beliefs** and the likelihood to give a **new probability distribution**, representing our new, **posterior beliefs** about the parameter's value.

CONFIDENCE INTERVALS

Bayesian inference allows for a more fluid interpretation of data, rather than the binary yes/no outcome of a **p-value significance test**.

The **posterior probability distribution** for *H* allows a scientist to ask "what's the **most likely range of values** for *H* to lie in?" Constraints such as this are known as **confidence intervals**.

For example, from the distribution, you can calculate what range of *H* encompasses, say, 95 percent of the probability. This would be a 95 percent confidence interval.

But confidence intervals can be wider (e.g., 99 percent) or narrower (e.g., 90 percent), depending on the problem at hand.

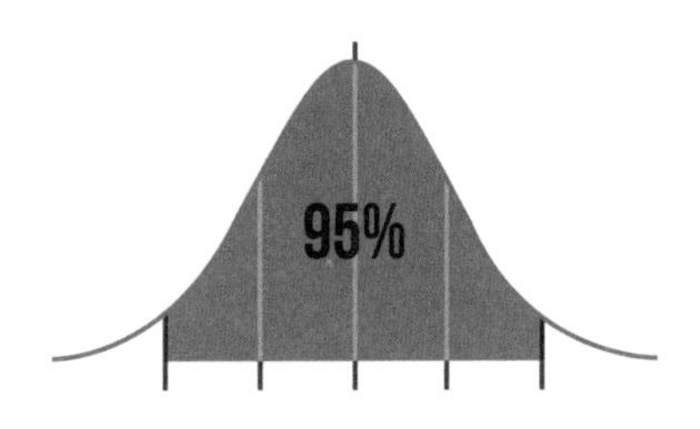

GAME THEORY

Game theory is quite literally a branch of math concerned with finding optimal strategies for playing games. It has applications in fields from poker, to politics and economics, to armed conflict.

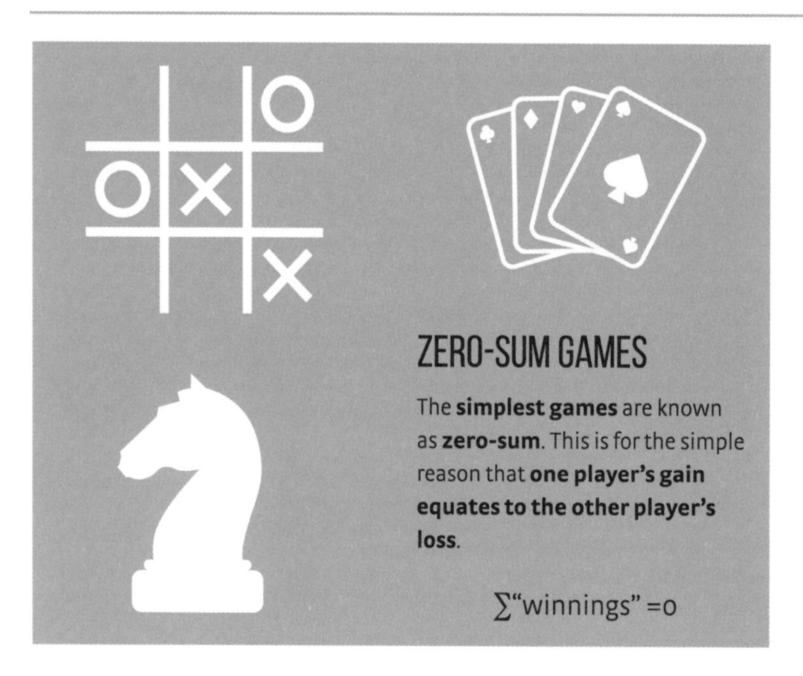

ZERO-SUM GAMES

The **simplest games** are known as **zero-sum**. This is for the simple reason that **one player's gain equates to the other player's loss.**

$$\sum\text{"winnings"} = 0$$

MINIMAX

In 1928, Hungarian-born mathematician **John von Neumann** proposed that the key to game theory must be what he called the **minimax theorem**. This says that a player's **optimal strategy is the one that yields the best outcome in the worst-case scenario**.

NASH EQUILIBRIUM

Minimax led directly to the concept of **Nash equilibrium**. This is a strategy that, **when played by all parties in a game**, leads to **a state where no one can improve their outcome by unilaterally changing strategy**. It's named after U.S. mathematician **John Forbes Nash**, who carried out much of the original work on the concept.

PRISONER'S DILEMMA

A classic example of **Nash equilibrium** comes from a problem known as the prisoner's dilemma. Two suspects are arrested for a crime. They're told that if neither **cooperate**, they'll both serve a year in prison. However, if **one betrays the other** then the betrayer goes free while the other does three years. And if they both betray each other then they both serve two years.

The so-called **pay-off matrix** for this "game," giving the **outcome for each player in every possible scenario**, looks like this:

B / A	B stays silent	B betrays
A stays silent	-1 / -1	0 / -3
A betrays	-3 / 0	-2 / -2

The Nash equilibrium is for both players to betray each other (the bottom-right cell in the pay-off matrix)—from here, neither player can improve their lot by changing strategy on their own.

TRAVELING SALESMAN PROBLEM

Imagine a salesman with a list of cities to call at. In what order should they visit the cities so that they minimize the total distance traveled? Answers on a postcard.

COMPLEXITY

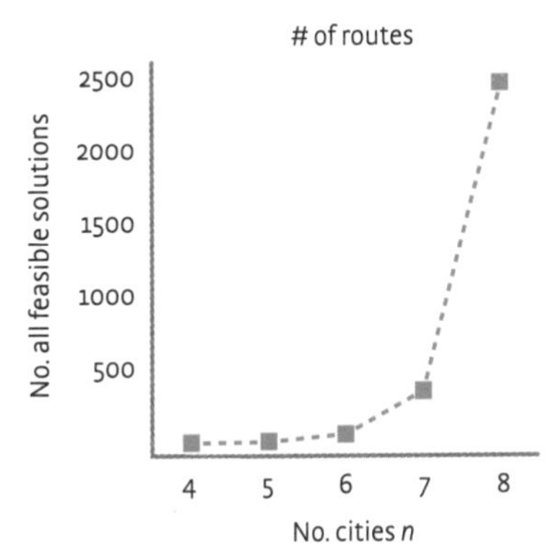

No general solution—that is, a solution that can be written down for an arbitrary number of cities—of the traveling salesman problem **is known**. Instead, **each case must be solved using numerical analysis and optimization techniques**. The problem is in fact **NP-hard**, meaning that the **time to find a solution grows disproportionately longer as the number of cities increases**.

BRANCH AND BOUND

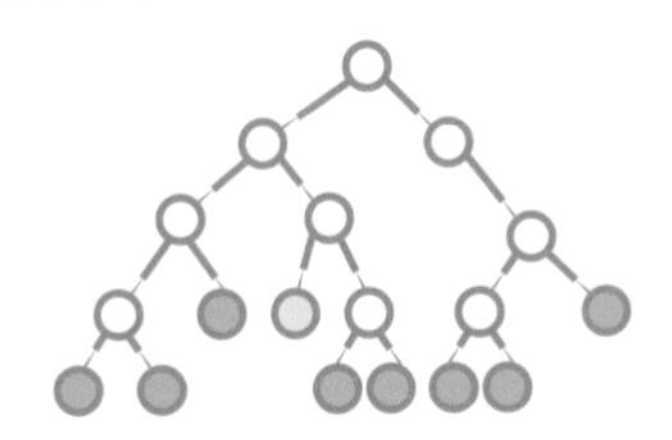

The most efficient **computational algorithm** to solve the problem is called **branch and bound**, developed by the British mathematician **Ailsa Land** and Australian **Alison Doig** in 1960.

The algorithm systematically **explores branches of a mathematical tree**—in this case, the tree of possible solutions to the problem—**estimating bounds on the shortest route** and immediately **discarding any routes that are longer**.

APPLICATIONS

As well as **route planning** (of the sort that might take place in your **GPS**), the traveling salesman problem also has applications in **manufacturing**, **scheduling**, **astronomy**, and **DNA sequencing**.

The problem has even been extended to include factors such as vehicle wear and bottlenecks, which make the shortest route not necessarily the quickest.

1832 The problem is first mentioned in a German handbook for traveling salesmen

1856 William Rowan Hamilton frames a similar problem mathematically, as an "Icosian game," minimizing distance traveled along the edges of an icosahedron

1960 Ailsa Land and Alison Doig propose the "branch and bound" algorithm

1972 American computer scientist Richard Karp proves that the problem is NP-hard

2004 The largest implementation of branch and bound solves the traveling salesman problem for 24,978 cities

RAMSEY THEORY

Devised by British mathematician Frank Plumpton Ramsey, this branch of graph theory deals with finding the conditions under which general statements in combinatorics hold true.

THE DINNER PARTY PROBLEM

A classic example is the dinner party problem. You're throwing a dinner party. To encourage conversation, **you want at least three guests to be mutual strangers, or to already be acquainted**. What's the **minimum number** to invite, to guarantee that one of these occurs?

THE SOLUTION

It turns out the number is **six**. To see why, draw six dots on a sheet of paper, each representing a guest. Now draw lines from each guest to every other guest—dark green if they know each other, light green if they don't.

With six guests (left-hand diagram) **it's impossible to connect them all without creating either a dark green triangle** (i.e., three people who know each other) or a light green triangle (three people who are strangers).

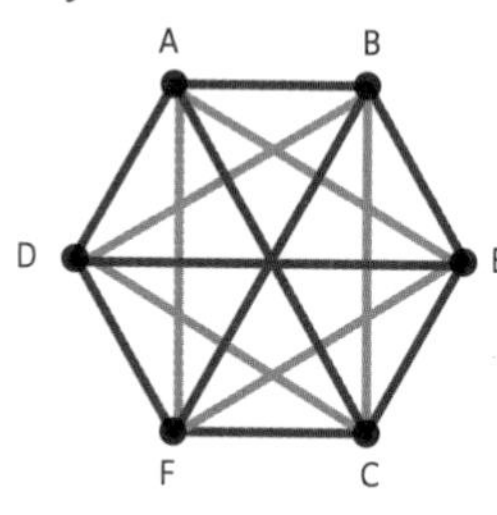

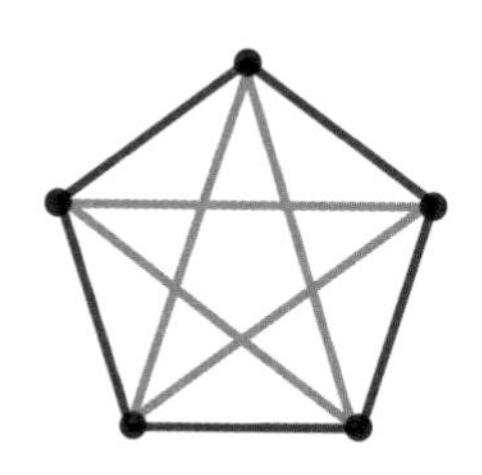

But this isn't the case with just five dots (right-hand diagram)—so six is the minimum number to invite.

RAMSEY NUMBERS

Ramsey generalized the dinner party problem to create the so-called **Ramsey numbers**, *R(m, n)*—the number of people to invite so that either *m* of them are already acquainted or *n* of them are not acquainted at all. The first few are:

m	*n*	*R(m, n)*
3	3	6
3	4	9
3	5	14
3	6	18
3	7	23
3	8	28
3	9	36
4	4	18
4	5	25

FRANK PLUMPTON RAMSEY

February 22, 1903 Ramsey is born in Cambridge. His father, Arthur Stanley Ramsey, is also a mathematician, and president of Magdalene College

1922 Ramsey learns German in order to translate Wittgenstein's *Tractatus Logico-Philosophicus*

1928 Publishes his work on Ramsey theory and the dinner party problem

January 19, 1930 Dies following surgery at Guy's Hospital, London, aged just twenty-six

JOHN VON NEUMANN

Few mathematicians have contributed more than von Neumann. A genius polymath, he carried out influential work in set theory, quantum mechanics, and the theory of games.

- **December 28, 1903** Von Neumann is born in Budapest, Austria-Hungary. A child prodigy, he can joke in Classical Greek and easily memorize and recite a page from the telephone book. He attends the Lutheran gymnasium, one of Budapest's foremost schools

- **1923** While still an undergraduate, von Neumann publishes the paper *The Introduction of Transfinite Ordinals*, which establishes the definition of an ordinal number as the set of all smaller ordinal numbers

- **1925** Von Neumann is awarded a degree in chemical engineering from the Swiss Federal Institute in Zurich and a Ph.D. in mathematics a year later from the University of Budapest

- **1928** He publishes the *Theory of Parlor Games*, based on poker and bluffing. This becomes an important work in game theory

- **1930s** Von Neumann teaches at the universities of Göttingen, Berlin, and Hamburg. In 1932, he publishes his influential work, *The Mathematical Foundations of Quantum Mechanics*

- **1933** Hitler comes to power and, like many other Jewish academics, von Neumann has to leave Germany. He becomes a professor of mathematics at the Institute for Advanced Study in Princeton, New Jersey

- **1942–46** During the Second World War, von Neumann works on the U.S. Manhattan Project, to produce the world's first atomic bomb

- **1956** He wins the inaugural Enrico Fermi Award, which is given for work on the development, use, or production of energy

- **February 8, 1957** Von Neumann dies of cancer in Washington D.C. He was remembered as a great wit at Princeton and was said to have played practical jokes on Einstein

GÖDEL'S INCOMPLETENESS THEOREM

Formulated by Hungarian mathematician Kurt Gödel, this is actually two theorems, asserting that any mathematical system contains statements that can be neither proved nor disproved.

THE FIRST THEOREM

The first incompleteness theorem states: **Any consistent formal arithmetic system *T* is incomplete. That is, there are statements concerning *T* which can be neither proved nor disproved within *T***. It means that there are truths within the mathematical system *T* that cannot be confirmed or refuted just by using the mathematical rules contained in *T*.

THE SECOND THEOREM

The second of Gödel's theorems follows from the first. It says that **no formal arithmetic system can prove its own consistency**. That is, the consistency of an arithmetic system *T* cannot be proven within the arithmetic framework provided by *T*.

Both theorems were published in Gödel's landmark 1931 paper **"On Formally Undecidable Propositions in Principia Mathematica and Related Systems I."**

EPIMENIDES PARADOX

The incompleteness theorems can be visualized to some extent as a **mathematical version of the Epimenides**, or **liar's paradox**. The paradox is encapsulated by the self-referential sentence: "this statement is false." **Either the statement is true, in which case it must be false, or it is false, in which case it has to be true**. In other words, whether or not the statement is true cannot be proven.

WHAT DOES IT MEAN?

As well as its implications for the **integrity of mathematics**, Gödel's incompleteness theorem has consequences in other fields.

In 1936, British computer scientist **Alan Turing** showed that it's **impossible to construct an algorithm that can determine whether an arbitrary computer program will eventually finish**, or will run forever. This became known as the **halting problem**.

And some physicists have claimed that Gödel's incompleteness theorem could also **rule out the possibility of formulating a theory of everything**—a unified theoretical description of all the phenomena in the physical world.

ALAN TURING

Renowned as the "father of theoretical computer science," Alan Mathison Turing's genius shone during the Second World War, when he created devices that cracked German military codes.

June 23, 1912 Born in London

1934 Graduates from the University of Cambridge, King's College, with highest honors in mathematics. Gains his Ph.D. at Princeton in the U.S.

1938 Begins working on the German Enigma ciphering system, at the Government Code and Cypher School (GC&CS)

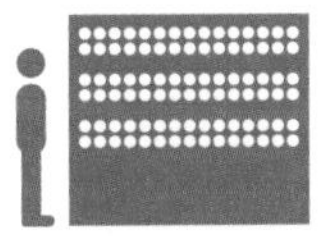

September 4, 1939 Begins work at Bletchley Park, Buckinghamshire, the wartime headquarters of GC&CS. Refines the specification of the "Bombe," an electro-mechanical machine created to crack the Enigma

1941 By this point, Turing and his team in Hut 8 at Bletchley Park are able to read naval Enigma messages. As a result, Allied convoys and warships can be warned of impending U-boat attacks

1944 A technique created by Turing, which he dubbed "Turingery," used for figuring out the rotor settings of the more complex Lorenz machine, is implemented on the Colossus computer; it enables GC&CS to monitor messages from German High Command

1945 Turing develops a speech-scrambling device, which he names "Delilah." He is awarded an O.B.E. that year, in recognition of his service

1946 After the war, he experiments with computer design and he presents a model for the Automatic Computing Engine, a fully electronic programmable computer

1952 Convicted of committing "gross indecency" with a man (homosexuality was illegal in Britain then). He agrees to hormone treatment to reduce libido

June 8, 1954 Turing was found dead at his home in Wilmslow, Cheshire. It's rumored that he committed suicide by taking cyanide. In 2013, he was given a royal pardon

BENFORD'S LAW

In a numerical dataset, you might expect the leading digits of each number to be evenly distributed. In fact, lower-value digits are much more likely than higher digits.

WHAT IS IT?

The phenomenon was first seen by American astronomer **Simon Newcombe** in the 1880s. It was rediscovered and popularized in 1938 by the physicist and engineer **Frank Benford**.

He found that the **leading digits of numbers in a real-life dataset are distributed as follows**:

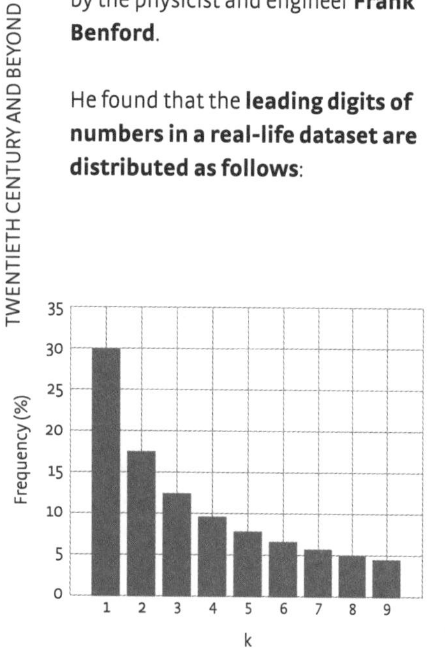

So, the leading digit is 30 percent likely to be a "1" but less than 5 percent likely to be a "9."

Benford tested the law on many datasets, including **constants of nature**, **population statistics**, **life expectancies**, and the **sizes of rivers**.

FORMULA

Benford's law can be **framed as a formula**. The leading digit d, for numbers expressed in base b, occurs with probability

$$P(d) = \log_b\left(1 + \frac{1}{d}\right)$$

The law is **most pronounced in large datasets containing numbers that span several orders of magnitude**. At the time of writing, **a full explanation has yet to be found**.

FRAUD DETECTION

This stark **contrast between a real-life set of numbers**, and **one that's been invented or generated randomly**, has found an application in **spotting fraudsters**. The distribution of digits in a fabricated table of accounts, for example, generally will not follow the pattern of Benford's law.

RELATED EFFECTS

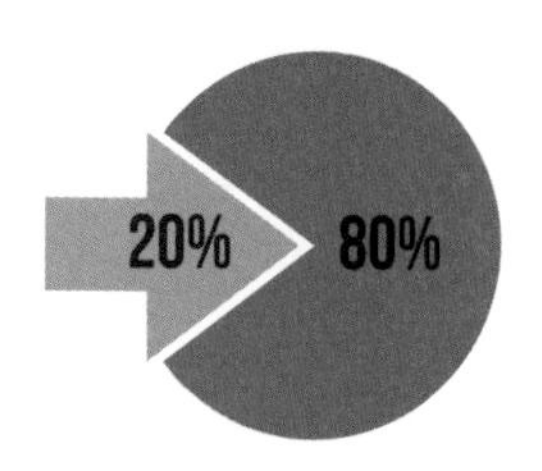

The **Pareto principle**: also known as the **eighty–twenty rule**, this states that **80 percent of monetary wealth is typically owned by 20 percent of the population**.

The

Zipf's law: the **frequency of a word in a text is inversely proportional to its rank**. So the *n*th most common word will occur with a probability proportional to 1/*n*.

Price's law: half of the publications in a given field are written by the square root of all authors in that field. So if there are a hundred authors, half of all the literature will be written by ten of them.

OPTIMIZATION

When people apply mathematical reasoning to the real world, they are often looking for the best deal—by minimizing cost or maximizing benefit. Welcome to the world of optimization.

WHAT IS IT?

Optimization is a technique that concerns finding **parameter choices for a mathematical function that either maximize or minimize its value**.

For example, let's say we want to minimize the function

$$f(x) = x^2 - 4x + 5$$

Plotting a few values is a good way to get a grip on the function's **behavior** (see right).

And from this, we can see straight away that the minimum value of the function is 1, and this occurs when $x = 2$.

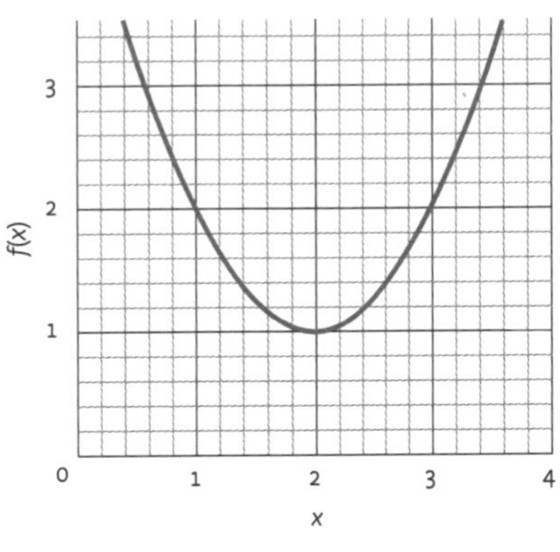

THE REAL WORLD

Real optimization problems are rarely this simple. Imagine instead if f depends on twenty different variables rather than one, and each variable can only take a certain permitted range of values.

Solving **constrained optimization problems** such as this typically requires **numerical analysis using specialist optimization algorithms**. This is why the field of optimization is sometimes said to straddle both **mathematics** and **computer science**.

THE DIET PROBLEM

One of the first real-life optimization challenges was the diet problem. Posed by the **U.S. Army** in the 1930s, the brief was to find the **cheapest combination of seventy-seven available foods to feed a soldier so that their recommended intake of nine key nutrients would be satisfied**.

It amounted to a constrained optimization problem in seventy-seven variables. In 1939, the economist **George Stigler** arrived at an **approximate solution** that came to $39.93 per soldier per year.

In 1947, the problem was solved rigorously using the ***simplex* optimization algorithm**—although, amazingly, the exact solution cost only $0.24 less than Stigler's estimate.

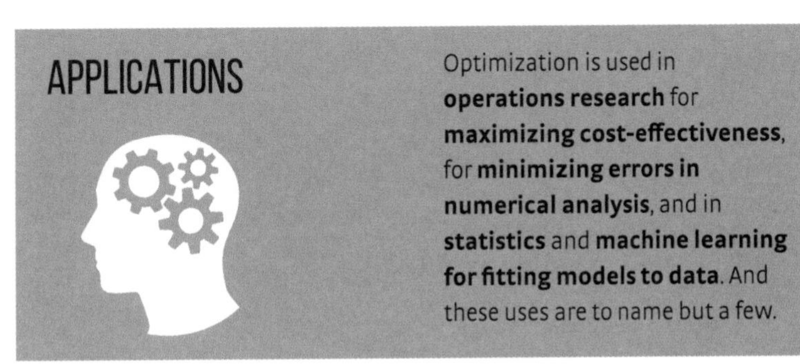

APPLICATIONS

Optimization is used in **operations research** for **maximizing cost-effectiveness**, for **minimizing errors in numerical analysis**, and in **statistics** and **machine learning for fitting models to data**. And these uses are to name but a few.

KURT GÖDEL

Famed for his "incompleteness theorem"—that there exist unprovable mathematical truths—Kurt Gödel was a logician who applied philosophy to mathematics, with revolutionary results.

EARLY YEARS

Gödel was born on **April 28, 1906, in Brünn, Austria-Hungary**. He was a sickly child, who suffered from **rheumatic fever**. He held a lifelong belief that the illness had affected his health.

Gödel studied at the **University of Vienna**, where he earned a **doctorate in mathematics** in 1929. He became **fascinated by mathematical logic** and was part of the **"Vienna Circle,"** an influential network of **scientists**, **philosophers**, and **mathematicians**.

INCOMPLETENESS THEOREM

In 1931, Gödel published one of science's most important findings—the "incompleteness theorem," which stated that **all logical systems have contradictions or tenets that can't be proven**.

PHILOSOPHY OF MATHEMATICS

From the early 1940s onward, Gödel spent most of his time contemplating the philosophical issues surrounding mathematics. He studied the works of philosophers such as Leibniz and Kant, published a number of research papers, and authored a book—*Is Mathematics a Syntax of Language?*—though it was, sadly, never published.

NAZI EXPANSION

After Hitler annexed Austria in 1938, **Gödel fled to Princeton**, where he took up a position at the **Institute for Advanced Studies**. He **befriended Einstein** and created a **mathematical solution for general relativity**—the **rotating "Gödel universe"**—that could allow for the possibility of **time travel**.

MENTAL HEALTH PROBLEMS

Gödel's **mental health declined** in later life and it is believed that he **starved himself to death** in 1978.

ARTIFICIAL NEURAL NETWORKS

Artificial neural networks are machine learning algorithms that mimic the connections between biological neurons in the brain. They are used to find hidden patterns in data.

1943 American scientists Warren McCulloch and Walter Pitts build the first mathematical model for a neural network

1969 American cognitive scientist Marvin Minsky and colleagues find the networks to be severely limited by the computing power of the day

1975 American researcher Paul Werbos develops the **backpropagation** algorithm for efficiently training neural nets

2012 Deep neural networks, with many hidden layers, begin to significantly outperform other machine learning techniques

HOW DO THEY WORK?

Neural networks consist of **multiple layers of simulated neurons**, the individual states of which are **stored in a computer**.

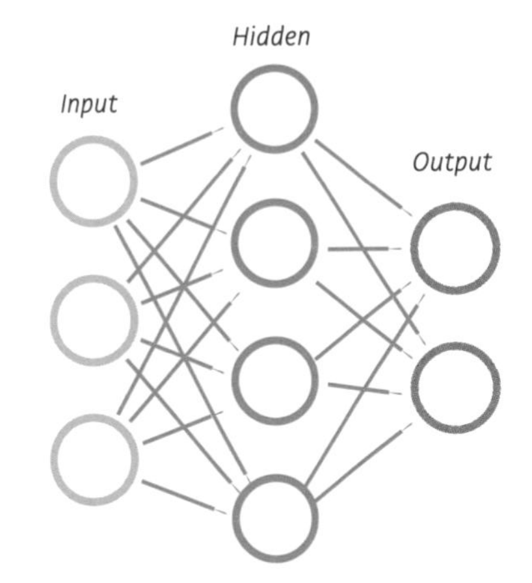

The first layer feeds directly from the **input data**. If the sum of each neuron's inputs exceeds a threshold (determined by **training**) then the output from that neuron is 1, otherwise it's 0.

In their simplest form, **each layer feeds its outputs forward to the next**. There may be more than one **hidden layer**, depending on the structure of the particular network.

Finally, **all the strands in the network come together to form the output**.

TRAINING

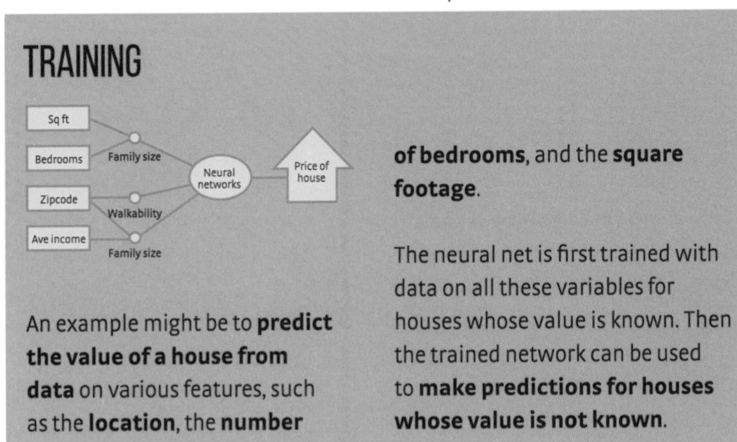

An example might be to **predict the value of a house from data** on various features, such as the **location**, the **number of bedrooms**, and the **square footage**.

The neural net is first trained with data on all these variables for houses whose value is known. Then the trained network can be used to **make predictions for houses whose value is not known**.

APPLICATIONS

The applications of artificial neural networks are countless. A few examples below.

Medical diagnosis: they can **scan a patient's X-rays** looking for abnormalities.

Driverless cars: neural nets analyze live video to **keep the car on the road** and **check for threats**.

Finance: applications here include **prediction of market prices** and **automated fraud detection**.

INFORMATION THEORY

The mathematical theory of how information is stored, transmitted, and manipulated was developed in the 1940s by the U.S. mathematician and electronic engineer Claude Shannon.

WHAT IS IT?

Shannon took the **binary digit**, or **bit**, to be the **fundamental unit of information**.

He studied the transmission of bits in the presence of **noise**, and found a fundamental quantity was the so-called **information entropy of a bit**, $H = -\sum p \log p$, where p is the probability that the bit takes the value 1.

Broadly speaking, H is a measure of the uncertainty in the bit's value.

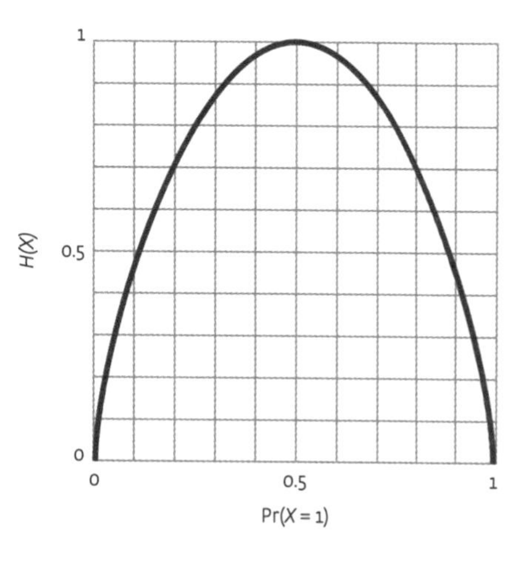

It **behaves in a similar way to entropy in thermodynamics**, hence the name.

APPLICATIONS

Applications of information theory include **data compression algorithms**, which enable you to store thousands of **songs on your phone**, and **error correction**—which helps supermarkets scan **barcodes**, even on a scrunched-up pack of chips.

THE KELLY CRITERION

In 1956, Bell Labs researcher **John Kelly** applied information theory to figure out the **optimum fraction of their bankroll a gambler should stake on a bet**. Stake nothing and they'll win nothing; stake everything and they'll face ruin.

Kelly found the optimal bet is:

$$f = \frac{p(b+1) - 1}{b}$$

where p is the **probability of the bet winning** and the **odds offered by a bookmaker** are b to 1.

For example, a gambler with $1,000, looking to back a 4-to-1 shot that has a 30 percent ($p = 0.3$) chance of winning, should stake $125.

CLAUDE SHANNON

April 30, 1916 Claude Elwood Shannon born in Petoskey, Michigan

1936 Graduates from the University of Michigan with degrees in electrical engineering and mathematics

1940 Receives his Ph.D. in analog computing, from MIT

1948 His seminal paper on information theory, "A Mathematical Theory of Communication," is published

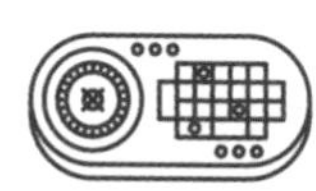

1961 Shannon and gambling math guru Edward Thorp build the world's first wearable computer—to beat roulette

February 24, 2001 Shannon dies in Medford, Massachusetts, following a struggle with Alzheimer's disease

ARROW'S IMPOSSIBILITY THEOREM

In 1951, American mathematician Kenneth Arrow proved that no electoral system that ranks candidates in order of preference can ever fairly represent the views of voters.

WHAT IS IT?

The theorem falls under the remit of **social choice theory**, a branch of **applied mathematics** that deals with collecting together opinions in order to **enact decisions** that are for the **greater good**.

FAIRNESS CRITERIA

Arrow defined three **fairness criteria** against which **electoral systems** could be gauged.

1. If all voters prefer option *X* to option *Y*, then the group should prefer *X* to *Y*.
2. If all voters' preferences between *X* and *Y* are unchanged, then the group's preferences between *X* and *Y* should also be unchanged.
3. **No single voter should ever possess the power to overturn the general preference of the group.**

PROOF

Arrow was able to show that an electoral system where voters **rank their preferences for three or more candidates is unable to satisfy these criteria**.

		Candidates		
		X	Y	Z
Voters	A	3	2	1
	B	1	3	2
	B	2	1	3

For example, in the diagram, voters *A*, *B*, and *C* express their preferences for options *X*, *Y*, and *Z*, where 3 is most preferred and 1 is the least preferred.

Voters *B* and *C* both rate *Z* higher than *X*. And yet the majority (two voters of three) prefer *X* to *Y* and *Y* to *Z*, and therefore by association *X* to *Z*.

KENNETH ARROW

August 23, 1921 Arrow born in New York City

1940 Graduates from the City College of New York with a degree in mathematics

1951 The impossibility theorem comes from results published in Arrow's Columbia University Ph.D. thesis

1972 Arrow shares the Nobel Prize in Economics

February 21, 2017 Dies in Palo Alto, California

JOAN CLARKE

An unsung heroine of the Second World War, Joan Clarke was one of the few female cryptanalysts at Bletchley Park, where the German military codes were broken.

June 24, 1917 Born in London. Attends Dulwich High School, where she wins a scholarship to Newnham College, Cambridge

1939 Gains double honors in mathematics although, under Cambridge University law at the time, she is prevented from receiving a full degree because of her gender

June 1940 Recruited to Bletchley Park by Gordon Welchman, her supervisor at Cambridge. Initially, she does clerical work because at the time women are not considered suitable for roles as codebreakers

Her talent shines out and she is invited to join the nerve center at Hut 8, where Alan Turing and his team are working on breaking the codes used by German U-boats, which were exacting heavy losses on Allied ships transporting troops and supplies across the Atlantic

1941–44 In an incredibly high-pressure environment, Clarke has to break the codes in real time, which often results in immediate military action

1944 Clarke becomes deputy head of Hut 8, though, due to her gender, she is paid less than male colleagues

After the war Clarke is awarded an MBE and works at Government Communications Headquarters (GCHQ). She shuns the limelight and only speaks reticently about her service when Bletchley Park becomes famous in the 1990s

1996 Clarke dies in Headington, Oxford. The Bletchley Park codebreakers are believed to have cut short the war by two years, saving millions of lives. She was among its stars

DID YOU KNOW?

Clarke and Turing became close during their time at Bletchley Park and were **engaged for a short time**. The pair **remained friends until Turing's death** in 1954.

EXTREME VALUE THEORY

Extreme value theory is a branch of statistics dealing with the likelihood of extreme events. Applications include the prediction of floods, stock market crashes, and sporting achievements.

WHAT IS IT?

The field of extreme value theory began in the early twentieth century with the work of English statistician **Leonard Tippett**. Commissioned to study the **strength of cotton thread**, Tippett constructed **probability distributions** for the strength of the **weakest cotton fibers**.

And this is what extreme value theory is all about: **predicting the distribution of the maxima and minima of quantities** over a **given time period**.

EXTREME VALUE DISTRIBUTIONS

Tippett's research was placed on a **more rigorous mathematical footing** in 1958 by the German mathematician **Emil Gumbel**.

The **Gumbel distribution** is a **continuous probability distribution** for a random variable x given by the formula:

$$f(x) = \frac{1}{\beta} e^{-(z+e^{-z})}; z = \frac{x - \mu}{\beta}$$

where μ and β are **parameters controlling the shape of the distribution**. Some examples are shown in the diagram:

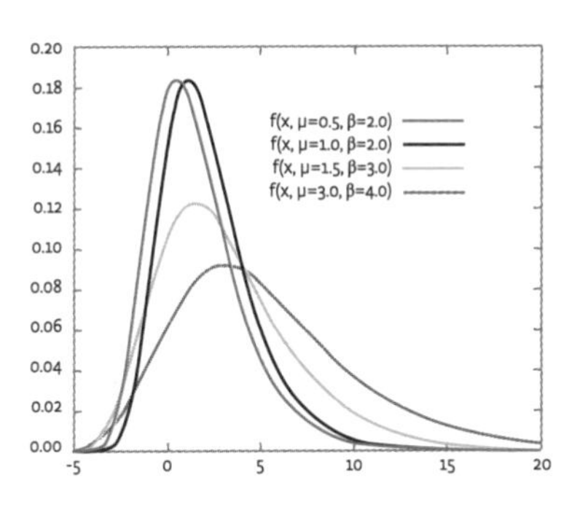

Gumbel showed that a variable obeying a **decaying exponential distribution** will have maxima approximated by the **Gumbel distribution**. It is one of many different extreme value distributions.

APPLICATIONS

Floods: the **maximum levels of rivers** and the **heights of freak waves** are all governed by extreme value distributions and can be used to **predict flood risk**.

Finance: extreme **highs and lows of the stock market**, or **large insurance losses**, also obey an extreme value distribution.

Traffic: predictions of **extreme road traffic** can be used to **manage congestion in city centers**.

Sports: according to extreme value theory, the **fastest possible time** in the 100-meters sprint is 9.51 seconds.

EMIL GUMBEL

- **July 18, 1891** Emil Julius Gumbel born in Munich, Germany
- **1932** Gumbel, a Jew, is forced out of Germany by the rise of Nazism
- **1958** Publishes his work on extreme value theory in the book *Statistics of Extremes*
- **September 10, 1966** Dies in New York City

OLGA LADYZHENSKAYA

Living under the Soviet regime for most of her life, Ladyzhenskaya overcame personal tragedy and political hurdles to make important contributions in fluid dynamics and differential equations.

HEROINE OF OUR TIME

Olga Aleksandrovna Ladyzhenskaya was **born in Kologriv**, a small town in Russia, on **March 7, 1922**.

Her father, **Aleksandr Ladyzhenski**, was a teacher who **inspired a love of mathematics in his daughter**. In 1937, he was **arrested** for being "an enemy of the people" and **executed**. The family was **thrown into poverty**.

Due to her father's status, Ladyzhenskaya was not admitted to Leningrad State University; however, she entered **Moscow State University** in 1943 and **later completed two Ph.D.s**.

Ladyzhenskaya **taught at Leningrad State University** and the **Steklov Institute, where she became head of mathematical physics**.

Always an independent spirit, Ladyzhenskaya **didn't shy away from criticizing the totalitarian state** and this placed her in a precarious position.

She carried out extensive research into the **Navier–Stokes equations**, a set of **partial differential equations** developed in the nineteenth century to explain **fluid dynamics**. Her discoveries in this field led to **important practical applications** in **oceanography**, **aerodynamics**, and **meteorology**, such as the **measurement of storm cloud movement**.

In the course of her long career, she **wrote in excess of 200 scientific papers and six monographs**.

In 1958, Ladyzhenskaya was **shortlisted for the Fields Medal**, one of the highest honors a mathematician can receive.

In 2002, she **won the Lomonosov Gold Medal for outstanding achievement in mathematics**.

She **died** in St. Petersburg, Russia, on **January 12, 2004**.

SMALL WORLD NETWORKS

Small world networks are a branch of graph theory, which supposes that large groups can be linked by a relatively small number of connections. The Kevin Bacon game is a well-known example.

WHAT ARE THEY?

Imagine you drew a dot on a sheet of paper for **every human being on Earth**. (There are over 7.5 billion of them, so you'd need a very large sheet of paper). Now **draw lines connecting the dots corresponding to people who know each other**.

So, for example, your dot on the paper would be connected in one step to all of your immediate acquaintances. Friends of friends would be separated by two steps, friends of friends of friends by three steps, and so on.

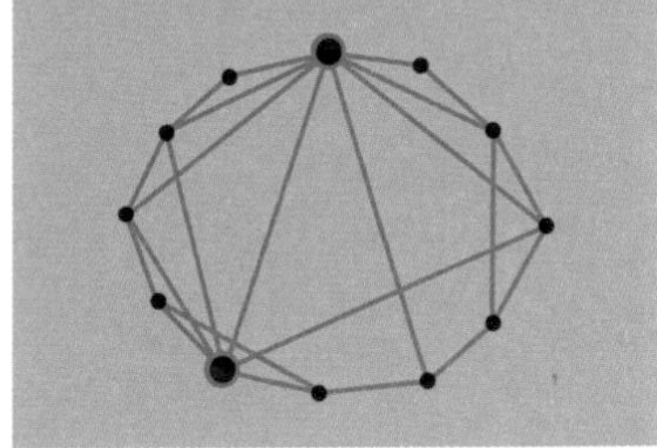

SIX DEGREES OF SEPARATION

Small world networks are subsets of dots that can be reached in a small number of steps from any starting dot.

For instance, the **"six degrees of separation" theory** supposes that **any two people on Earth can be connected by their acquaintances in no more than six steps**.

Typically, the number of steps, L, to connect two random dots, or **nodes**, scales with the **logarithm** of the total number of nodes, N. That is,

$$L = k \log N$$

Where k is a constant. So increasing the size of a network from 10 to 100 will require double the number of steps to connect everyone.

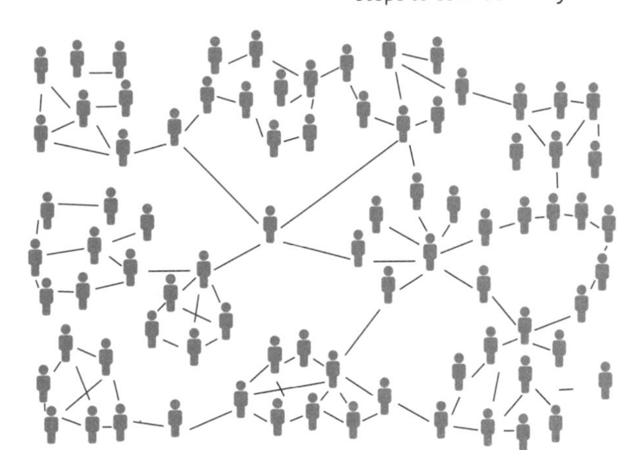

THE KEVIN BACON GAME

The theory of small world networks has also been applied to **movies**, leading to what's become known as the Kevin Bacon game.

How many steps does it take to connect an actor to Kevin Bacon, based on **movies they've co-starred in?** For example, Harrison Ford has a Bacon number of 2—he was in *42* with Brett Cullen, who was in *Apollo 13* with Kevin Bacon.

Similarly, there are the **Erdös numbers—linking scientists by co-authorship of research papers with Hungarian mathematician Paul Erdös**.

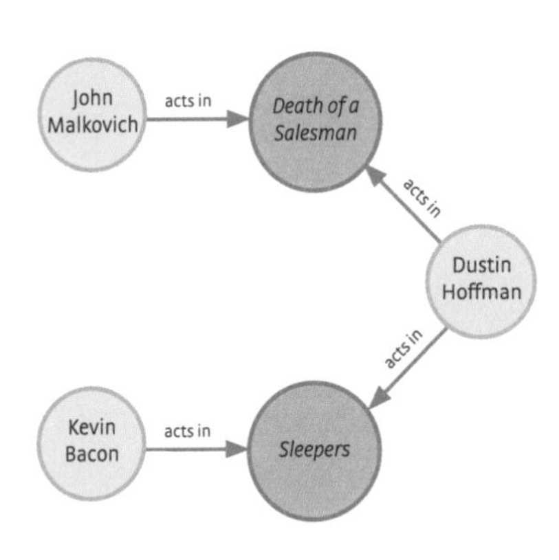

CELLULAR AUTOMATA

Cellular automata are virtual entities that demonstrate how complex behaviors can emerge from relatively simple sets of rules. They were implemented in the Game of Life computer model.

GAME OF LIFE

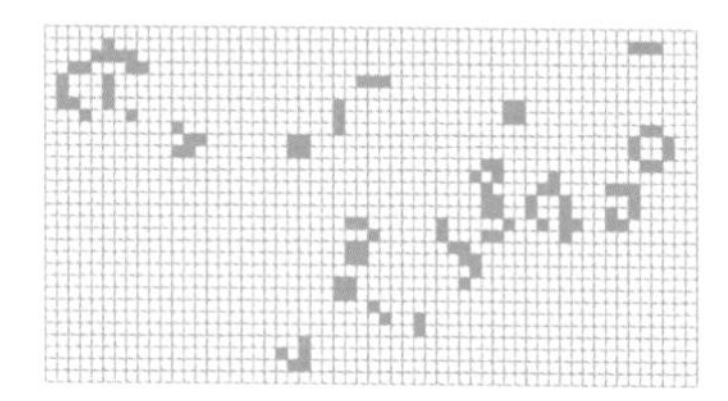

The most famous example of cellular automata is **Conway's Game of Life**, developed in the early 1970s by British mathematician John Conway and popularized by the science writer **Martin Gardner**.

Conway's model amounted to a **grid of squared paper** with **some cells left blank** and **others shaded in. A set of rules dictated how any given pattern of cells evolved from one time step to the next.**

THE RULES

For example, a shaded square with less than two shaded neighbors **"dies"** and becomes a white square. While a white square with three shaded neighbors **"comes alive,"** becoming shaded at the next step.

When the system was switched on and left to run, **incredibly diverse and complex behaviors emerged**, including **static and oscillating patterns**, **traveling shapes**, and **self-perpetuating cycles of replication and death**.

Before After

Loneliness
A cell with less than two neighbors dies

Before After

Reproduction
An empty cell with exactly three neighbors becomes alive

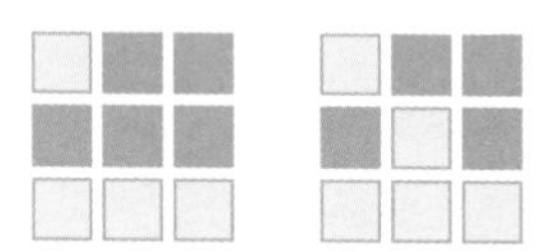

Overcrowding
A cell with more than three neighbors dies

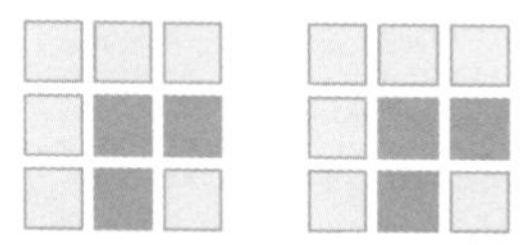

Stasis
A live cell with two or three neighbors remains the same

A NEW KIND OF SCIENCE

In 2002, British mathematician and computer scientist **Stephen Wolfram** published his book ***A New Kind of Science***, in which he set out his thesis that cellular automata are of **fundamental importance to the workings of the physical world**.

Wolfram argues that rather than simply being a tool to aid our understanding of science, **computation**—as exemplified by the **simple computer programs** that govern cellular automata—**is itself a component of nature**.

- **1948** John von Neumann writes a paper called "The General and Logical Theory of Automata"
- **1969** American computer scientist Alvy Ray Smith develops the first mathematical theory of cellular automata
- **1970** Mathematician John Conway devises the Game of Life
- **2002** Scientist Stephen Wolfram publishes his book *A New Kind of Science*

NONTRANSITIVE DICE

Everybody knows that gambling is for fools, right? Not if you're playing with a set of nontransitive dice and happen to know what you're doing.

WHAT ARE THEY?

Imagine you have three dice labeled *A*, *B*, and *C*. If *A* rolls a higher score than *B* more often than not, and *B* rolls a higher score than *C* more often than not, then **intuition might lead you to believe that *A* will roll higher than *C* more often than not.**

However, **for nontransitive dice the complete opposite is true** and *C* will, **in the long run**, beat *A*.

WHAT'S GOING ON?

Imagine the three dice are all six-sided, **but rather than the usual numbers 1 to 6 on each face, they are numbered as follows**:

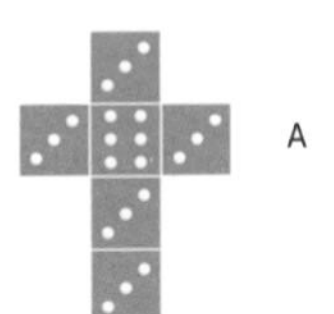

So when *A* plays *B*, *A* will roll a six one-sixth of the time, which *B* can't beat. The remaining five-sixths of the time, *A* rolls a three, which wins whenever *B* rolls a two—and this happens with **probability 3/6**. So *A* then wins with probability

$$p = \frac{1}{6} + \frac{5}{6} \times \frac{3}{6} = \frac{6+15}{36} = \frac{21}{36} = \frac{7}{12}$$

This is bigger than **1/2**. Similarly, you can work out the probability that *B* beats *C*; it's also **7/12**.

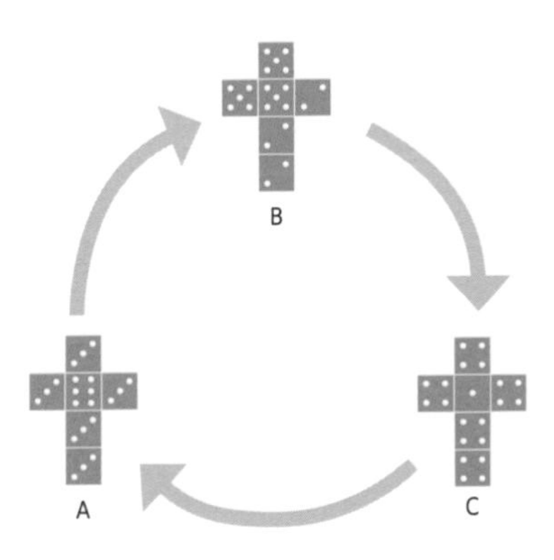

But then, for *C* to beat *A*, *C* must roll a four (probability 5/6) and *A* must roll a three (also probability 5/6). And so *C* wins with probability

$$p = \frac{5}{6} \times \frac{5}{6} = \frac{25}{36}$$

And this is also bigger than ½, meaning that *C* beats *A* more often than not, even though *A* beats *B* and *B* beats *C*.

DOUBLE DOWN

$$$

And so the **smart gambler graciously allows their opponent to select their die first, and then from the remaining two, picks the one which they know is most likely to win**. If nothing else, nontransitive dice illustrate **what a slippery animal probability really is**.

KATHERINE JOHNSON

This remarkable American mathematician overcame racial and gender prejudice to work at NASA, where she calculated the orbital mechanics of historic missions, including Apollo 11.

Katherine Johnson (**née Coleman**) was **born on August 26, 1918,** in **White Sulphur Springs, West Virginia**.

Her mathematical skill shone from an **early age**. She entered **West Virginia State College** at the age of fourteen and graduated five years later with **highest-honor degrees in mathematics and French**.

Johnson was **handpicked** to become **one of the first three African-American students** to enroll in a **graduate mathematics program** at **West Virginia University**. She **quit the course after a year to raise a family**.

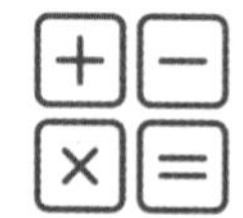

In 1953, she was **recruited to the National Advisory Committee for Aeronautics (NACA) West Area Computing unit**. This comprised a group of **African-American women who excelled in mathematics and carried out complex calculations for engineers**. Their work was vital to the success of the US space program. NACA was **segregated**, however, and the **women had to work, eat, and use bathrooms separate to white colleagues**.

NACA became NASA and Johnson joined its **Space Task Group**, which **managed the crew launches**.

In 1962, **John Glenn** was preparing for America's first orbital space flight. **Computers processed and determined the trajectory**. However, Glenn distrusted the machines and asked that Johnson check the calculations by hand. She recalled Glenn saying, "If she says they're good, then I'm ready to go." **The mission was a success**.

Johnson helped to calculate the **space trajectories for *Apollo 11***, which landed the **first men on the Moon**. She also worked on the **Space Shuttle missions** during her thirty-five-year career at NASA.

In 2015, **President Obama awarded her the Presidential Medal of Freedom**, America's **highest civilian honor**.

The film ***Hidden Figures*** (2016) **dramatized the lives of Katherine Johnson and her colleagues Dorothy Vaughan and Mary Jackson at NASA**.

BENOIT MANDELBROT

Mandelbrot inspired the world with his beautiful computer-generated images of fractals and helped us to understand their importance in nature. He pioneered a whole new field of math.

November 20, 1924 Born in Poland

1936 His family being Jewish, they flee to France to avoid Nazi persecution, where they live undercover and in constant fear

1952 Having previously studied at the École Polytechnique in Paris and the California Institute of Technology (Caltech), Mandelbrot completes a doctorate in mathematical sciences at the University of Paris

1958 Recruited to join IBM's Thomas J. Watson Research Center in New York, where he remains for the next thirty-five years, toward the end of his career becoming Sterling Professor of Mathematical Sciences at Yale University

October 14, 2010 After having won many awards during his career, including Israel's Wolf Prize for physics and being appointed an Officer of France's Légion d'Honneur, Mandelbrot dies in Cambridge, Massachusetts

THE WONDER OF FRACTALS

Mandelbrot was fascinated by **fractals—shapes that have rough outer edges, formed by the repeated iteration of an often simple rule**. Using the **computer wizardry** at his disposal at **IBM**, he created astonishing and beautiful fractal images using **computer code**. He wrote the landmark book ***The Fractal Geometry of Nature*** (1982) and **established fractals as a respected area of mathematical study**.

MANDELBROT SET

This famous fractal was named after Mandelbrot. It's key in **chaos theory**—where **slight changes** in initial conditions can produce **dramatic shifts** in the behavior of a system. Mandelbrot saw fractals as a way to **generate apparent complexity from simple rules**, and believed that this is how **seemingly complex** or **"rough" forms in nature**, such as rocky shorelines, could arise.

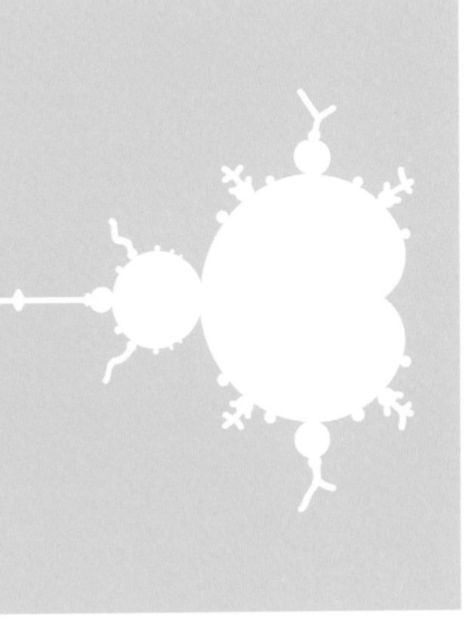

P VS NP

If a candidate solution to a problem can be checked quickly, does that mean you can solve the problem from scratch quickly too? Sounds a simple enough question, but it's not. This is P vs NP.

POLYNOMIAL TIME

At the heart of the P vs NP problem is a mathematical concept known as **polynomial time**.

A **polynomial** in some variable *x* **is a sum of terms involving x raised to various powers**. For example, $x + 2x^2 - x^3$ is a polynomial. Contrast that to an **exponential, where x is itself the power**, for example e^x.

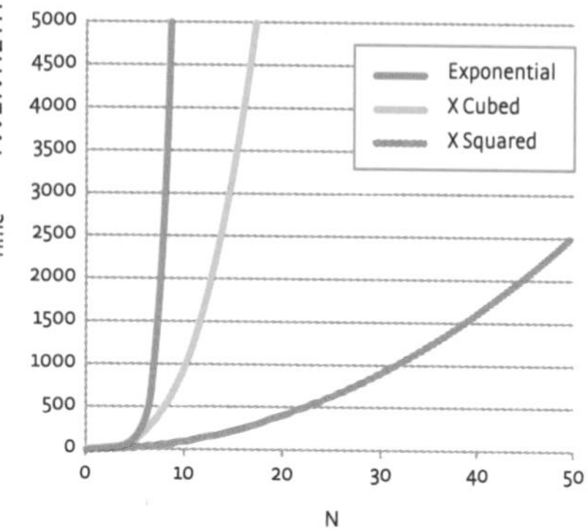

Polynomial time means that the time to compute **the answer to a problem grows as a polynomial function of its complexity**. This means the **growth is slow**, and the **problem is relatively quick to solve**.

Compare that to the other alternative: **exponential time**, where the time to solve the problem grows as an exponential function of its complexity, and rapidly becomes large.

SO WHAT IS P VS NP?

In **complexity theory**—the branch of **computer science** that classifies problems by their **level of difficulty**—a **problem that can be solved in polynomial time** ranks as "P." The **class of problems for which a solution can be verified in polynomial time** is called "NP," short for **"nondeterministic polynomial time."**

If P = NP, it **would mean that problems verifiable in polynomial time were also solvable in polynomial time**.

Whether this is the case remains an open question, and there's **$1 million** up for grabs from the **Clay Mathematics Institute** for anyone who can provide a solution.

1955 John Nash speculates that some problems in cryptography may have exponential complexity

1971 American-Canadian computer scientist Stephen Cook makes the first statement of the P vs NP problem

1979 Soviet mathematician Leonid Khachiyan shows that linear optimization is both NP and P

$$$

2000 P vs NP is selected by the Clay Mathematics Institute as one of its Millennium Prize Problems

CHAOS THEORY

Chaos is the emergence of apparently unpredictable behavior from seemingly well-behaved laws of math and physics. It is characterized by extreme sensitivity to initial conditions.

WHAT IS IT?

Chaos is **everywhere in the physical world**, from a **dripping tap** to the **tumbling moons of planets**.

In the 1960s, American mathematician **Edward Lorenz** was running **computer simulations of the weather**. He noticed that entering the initial conditions for each simulation to **six decimal places** (e.g., 0.387129) yielded **results** that were **dramatically different** from when he truncated the inputs to just **three decimal places** (0.387).

THE BUTTERFLY EFFECT

Lorenz's effect **wasn't random**—the system **evolved deterministically from each given starting state**. But he found that **tiny changes in the starting state led to enormous changes in the output**.

Lorenz came up with a metaphor to describe it—the **butterfly effect**—speculating that the **equations describing the weather are so sensitive that a butterfly beating its wings on one side of the world may influence the path of a tornado on the other**.

The effect was later named **chaos**.

CHAOTIC ATTRACTORS

The behavior of many physical systems tends toward states called **attractors**. For example, if you **plot position against speed for a frictionless swinging pendulum**, the **pendulum's motion will trace out a circle**.

Chaotic systems typically have attractors that are **fractal** in nature. When Lorenz plotted out the behavior of his weather model, he found a **beautiful fractal figure-eight pattern**, now known as the **Lorenz Attractor**.

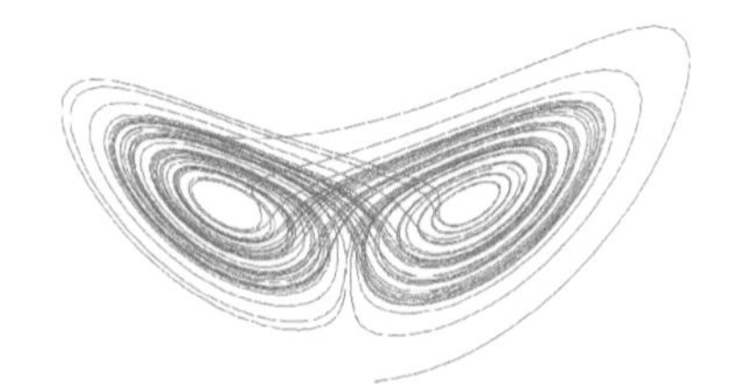

1961 Edward Lorenz notices sensitivity to initial conditions in his weather simulations

1972 Lorenz gives a talk entitled "Does the flap of a butterfly's wings in Brazil set off a tornado in Texas?", thus coining the term butterfly effect

1975 James Yorke and Tien-Yien Li first use the name chaos

1987 James Gleick's book *Chaos: Making a New Science* is published, bringing chaos theory to a popular audience

PUBLIC-KEY CRYPTOGRAPHY

The trouble with conventional encryption is that you have to tell others the key. And, sooner or later, it falls into the wrong hands. Public-key cryptography is a way around that.

WHAT IS IT?

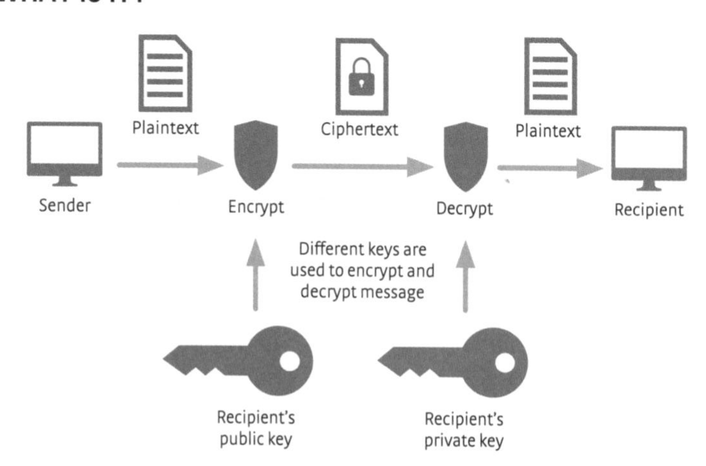

In public-key cryptography, there are actually **two keys**—one **public** and one **private**. The **private key is only needed to decrypt messages sent to you**, so you keep that to yourself. **Encrypting messages**, on the other hand, is **done with the public key**. Since you don't mind who's able to do this, you broadcast the public key to all and sundry.

It's rather **like sending out padlocks to everyone you know**. Anyone can put a message in a box and lock it—but **only you have the key needed to look inside**.

HOW DOES IT WORK?

The **algorithms** used in public-key encryption rely on the **difficulty in factorizing large numbers**—that is, splitting numbers into two smaller numbers which, when multiplied together, get you back to the original number. For example,

$$Z = X \times Y$$

Here, X and Y form the private key needed to decrypt messages. And they're then multiplied together to form the public key Z. But given just Z, it's very **hard to work backward** to get X and Y.

1970–74 Scientists at Britain's Government Communications Headquarters (GCHQ) carry out the first work on public-key cryptography, but the research is initially classified

1976 Americans Whitfield Diffie and Martin Hellman publish a protocol for exchanging cryptographic keys over a public channel

1977 The GCHQ work is independently replicated by MIT researchers Rivest, Shamir, and Adleman, and becomes known as RSA encryption

1994 American mathematician Peter Shor publishes a quantum algorithm for factorizing large numbers that could undermine RSA encryption

QUANTITATIVE FINANCE

The application of mathematical and statistical techniques to the world of finance and economics is known as quantitative finance. It has seen its share of success and failure.

1827 Scottish botanist Robert Brown observes random motion of particles inside pollen grains. It becomes known as Brownian motion, and its study is the basis for analyzing random walks, or stochastic processes, as they're also called

1863 French stockbroker's assistant Jules Augustin Frédéric Regnault is the first to suggest using a random walk to model the behavior of stock market prices

1900 French mathematician Louis Bachelier is the first to produce a rigorous model of the stock market using stochastic mathematics, in his Ph.D. thesis "The Theory of Speculation"

1906 Russian mathematician Andrey Markov develops the theory of random walks further

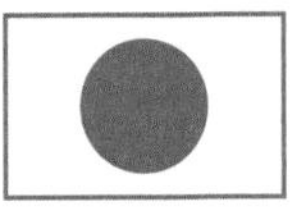

1942 Japanese mathematician Kiyosi Itô applies the rules of calculus to stochastic processes, enabling integration, and differentiation of random walks

1945 Austrian-born economist Friedrich Hayek argues that markets are an efficient way of aggregating economic information—an early application of the "wisdom of crowds" hypothesis

1952 American economist Harry Markowitz develops quantitative measures of correlations in the values of different stocks, enabling the construction of mathematical risk models

1965 American economist Paul Samuelson pioneers the application of Itô calculus to the financial markets

1969 Economists Fischer Black and Myron Scholes apply stochastic methods to derive a model for pricing financial options—essentially trading the right to buy or sell stock at a fixed price at some point in the future

1970 Eugene Fama, the "father of finance," formalizes Hayek's work as the efficient-market hypothesis

2008 Global financial markets are driven into meltdown as the U.S. subprime mortgage market collapses. Some of the blame is placed on the failure of quantitative models to accurately estimate the likelihood of rare events, or to account adequately for correlations

ROGER PENROSE

Renowned for his contributions to the mathematical physics of general relativity and black holes, algebraic geometer Roger Penrose was knighted for his services to science.

EARLY LIFE

Roger Penrose was **born in Colchester, Essex**, on **August 8, 1931**. He studied at **University College, London**, where he graduated with **highest honors in mathematics**. He completed his **Ph.D. at the University of Cambridge** on **tensor methods in algebraic geometry**.

Penrose has had a **long teaching career** at universities including **Birkbeck College, London**, and **Princeton**. He is **Emeritus Rouse Ball Professor of Mathematics at the University of Oxford**.

BLACK HOLES

Penrose collaborated with theoretical physicist **Stephen Hawking**. In 1969, they presented their **ground-breaking conjecture** that **all matter within a black hole collapses to a point, known as a singularity**, where **mass** is **compressed to infinite density**. Hawking and Penrose received **many prestigious awards for their findings**, including the **1988 Wolf Prize**.

Penrose developed what's become known as the **"Penrose diagram"** to illustrate **space and time within a black hole**.

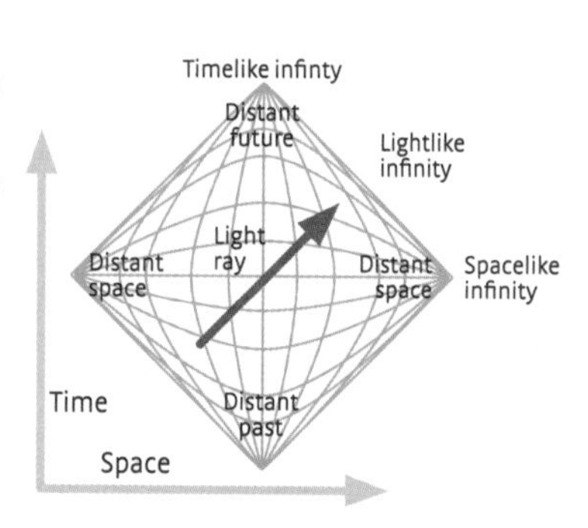

PENROSE TRIANGLE

In the 1950s, Penrose presented the **Penrose triangle**, an **optical illusion**, which he described as **"impossibility in its purest form."** The artist **M.C. Escher** used it in his famous work ***Waterfall***.

PENROSE TILING

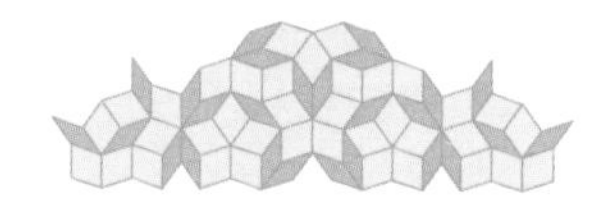

In 1974, Penrose became famous for the discovery of a **set of two tiles that could be used to tile an infinite plane only in a pattern that never repeats (aperiodicity)**. The shapes he used were **thick and thin rhombi** (diamond shapes). In **tiling theory**, it **remains to be proven whether a single tile can be used to cover a plane aperiodically**.

AWARDS

Penrose has won **many prestigious awards**, including the **2008 Copley Medal** for **"his beautiful and original insights into many areas of mathematics and mathematical physics."** He was **knighted in 1994**.

WALLPAPER GROUPS

Wallpaper groups are the subset of group theory responsible for classifying the symmetries of 2-D interlocking patterns. They're like tessellations on steroids.

WHAT ARE THEY?

There are **seventeen different wallpaper groups**, classifying 2-D patterns by their **translational**, **rotational**, and **reflectional symmetries**. They were discovered in 1891 by Russian mathematician **Evgraf Fedorov**.

The simplest is the group p1, which **consists purely of translations** (i.e., just **repeating a pattern without reflecting or rotating** it). For example,

On the other hand, the group p4 has **two 90-degree rotational symmetries** and one **180-degree rotational symmetry**, though **no reflections or translations**. For example,

SIMPLE TILING

Related to wallpaper groups is the concept of **tilings—ways to cover a 2-D plane using interlocking shapes**.

The **simplest kind** of tiling is **periodic**. If you tile a surface with **squares** or **regular hexagons**, then the **pattern repeats**—slide it around and you can make it match up with itself again.

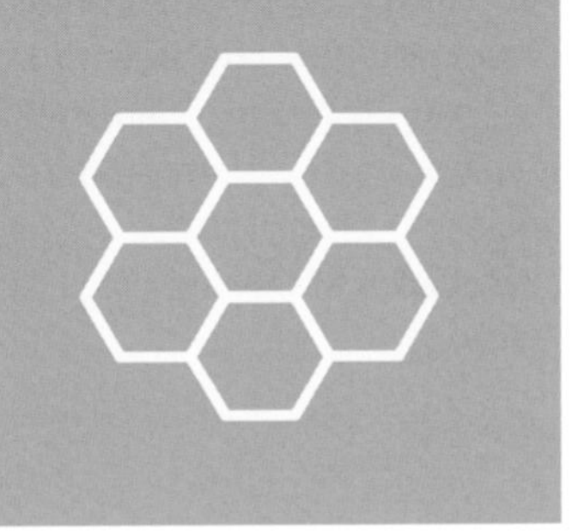

APERIODIC TILING

Far more interesting are **aperiodic tilings— where a) no matter how much you move, the pattern it never quite repeats, and b) there is no periodic way of arranging the tiles that form the pattern**. The simplest is a **radial pattern**.

In 1974, **Roger Penrose** found **two rhombus-shaped tiles that could form an aperiodic tiling** but that **couldn't be rearranged to tile the 2-D plane periodically**.

QUASICRYSTALS

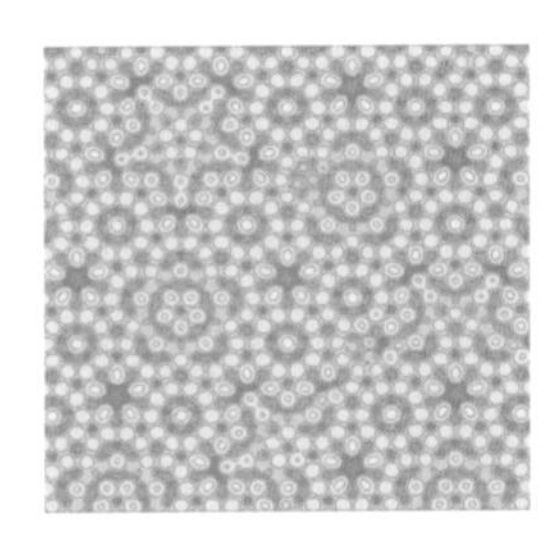

In 1982, Israeli materials scientist **Dan Shechtman** discovered a material—an **alloy of aluminum and manganese**—the **molecular structure** of which was **arranged into the 3-D analog of an aperiodic tiling**.

The structure was **neither amorphous** (where molecules are arranged haphazardly) **nor crystalline** (where molecules are arranged in a fixed, repeating geometric pattern). It **became known as a quasicrystal**.

JOHN NASH

A mathematical genius, Nash won the Nobel Prize in Economics for his masterful work in game theory. His struggle with schizophrenia was depicted in the 2001 film A Beautiful Mind.

LIFE

John Forbes Nash Jr was **born on June 13, 1928, in Bluefield, West Virginia**. His academic talent emerged early in life and **while at school he took advanced mathematics courses at a local college**.

Nash studied **chemical engineering** at **Carnegie Institute of Technology** before **switching to mathematics**. His Carnegie professor **Richard Duffin** recommended him to **Princeton University** for postgraduate study, stating that he was a "**mathematical genius**."

At Princeton, Nash was awarded a **Ph.D. in game theory**, which would become a lifelong fascination for him. This branch of mathematics is **applicable whenever there is any element of competition**, for instance in **wartime** or where there are **rival economic interests**.

Nash spent most of the 1950s researching at the **Massachusetts Institute of Technology**. He then began a long association with Princeton University.

BATTLING SCHIZOPHRENIA

During the late 1950s, Nash began to suffer bouts of **mental illness**, which manifested as **paranoia**. He was admitted to **hospital** in 1959 for **a year**. Nash battled **paranoid schizophrenia** for the next decade, with repeated spells in psychiatric hospitals. He **recovered gradually** and worked at Princeton when possible.

AWARDS AND ACCLAIM

In 1978, Nash was awarded the **John von Neumann Theory Prize** for **Nash equilibria—unexploitable strategies in game theory**.

The 1994 **Nobel Prize in Economics** was awarded to Nash, **John C. Harsanyi**, and **Reinhard Selten**, for their work on game theory.

In 2015, he shared the **Abel Prize** with **Louis Nirenberg** for the study of **partial differential equations**. Tragically, **Nash and his wife Alicia died in a car accident on May 23, 2015**, on their way home from the airport **after accepting the prize in Norway**.

EDWARD THORP

Casinos fear him, gamblers revere him. Ed Thorp is the professor who mathematically proved that you can "beat the dealer."

MONEY BALL

Edward Oakley Thorp was **born on August 14, 1932, in Chicago, Illinois**. He was awarded a **Ph.D. in mathematics from the University of California** and became a professor in 1961. He has **taught at New Mexico State University and the University of California, Irvine**.

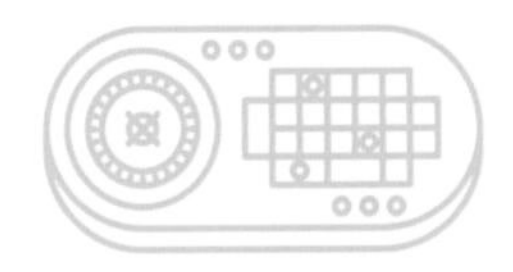

Thorp became interested in **gambling** as a teenager, when it occurred to him that a **roulette ball moves like a planet in orbit**. He **wondered if its motion might be predictable**.

His interest in gaming grew and he began visiting **casinos** in 1958, studying **blackjack tables** in particular. He realized that he could devise a **card-counting strategy to give a player an edge**.

Thorp was studying **mathematics** at **Massachusetts Institute of Technology (MIT)** at the time and **entered data into the faculty computer** (an IBM 704). He found that **card-counting** was **particularly effective toward the end of a deck** that was **not shuffled after each deal**.

BEAT THE DEALER

Thorp **tested his theory at casinos in Las Vegas, Reno, and Lake Tahoe**, using **funds provided by professional gambler Manny Kimmel**. They **won $11,000 in one weekend**. Thorp used **disguises**, such as false beards and dark glasses, during casino visits.

In 1966, Thorp wrote the **best-selling book *Beat the Dealer***, in which he explained his **strategy for card-counting**. **Casinos now shuffle blackjack cards before the end of the deck**, to counter such methods.

During the 1960s, Thorp began to apply his expertise in **statistics** and **probability to the stock market** and he **set up his own hedge fund**. His **net worth** is **now estimated to be $800 million**.

MICHAEL ATIYAH

Touted as Britain's greatest mathematician since Newton, Sir Michael Francis Atiyah was the driving force behind important developments in algebraic geometry and topology.

LIFE

Michael Francis Atiyah was **born on April 22, 1929 in London**. His father was Anglo-Lebanese and his mother was Scottish. He attended **schools in Khartoum**, **Sudan**, and **Cairo** before returning to the UK and studying at **Manchester Grammar School**.

From 1949 to 1955, Atiyah studied at **Trinity College, Cambridge**, where he was awarded a **doctorate**. His thesis title was **"Some Applications of Topological Methods in Algebraic Geometry."**

He enjoyed a long academic career and became **Savilian Professor of Geometry at the University of Oxford**, **master of Trinity College, Cambridge**, and **chancellor of the University of Leicester**.

SPECIALISM

Atiyah was a masterful **geometrist**. He made significant advances into the **connections between algebraic geometry and topology**, and his work in these areas earned him the **1966 Fields Medal**.

Alongside mathematicians **Alexandre Grothendieck** and **Friedrich Hirzebruch**, he pioneered the development of **K-theory**—a branch of math that **combines algebra**, **geometry**, and **number theory**.

Working with the U.S. mathematician **Isadore Singer**, Atiyah found a **link between the topological and analytical properties of the solutions to certain differential equations**. This became known as the **Atiyah–Singer index theorem** and they won the **2004 Abel Prize** for their findings. Atiyah also made contributions to **complex analysis**, **gauge theory**, and **superstring theory**.

AWARDS AND PUBLIC PROFILE

Atiyah was the **first director of the Isaac Newton Institute for Mathematical Sciences in Cambridge**, which was opened in 1992. He was also a **keen advocate for public engagement in mathematics** and gave popular, inspiring talks on the subject.

He received a **knighthood** in 1983 and won the **1988 Copley Medal** for contributions to **geometry**, **topology**, and **analysis**. Atiyah **died on January 11, 2019**, at the age of eighty-nine.

PAUL ERDÖS

An academic nomad, Erdös roamed the world seeking out collaborators who could help him solve some of the thorniest problems in twentieth-century mathematics.

Erdös is **born on March 26, 1913**, in **Budapest**, which was then part of Austria-Hungary. His **parents are both mathematics teachers**

His **father is held in a Russian prisoner-of-war camp** between 1914 and 1920 and his **mother works long hours**, so **Erdös passed the time by teaching himself mathematics**

From 1930, Erdös studies at the **University of Budapest**, where he gains a **doctorate in mathematics**. As a student, Erdös and his friends champion **Ramsey theory**, an area of graph theory

Erdös is Jewish and the rise of **anti-Semitism** in Hungary **prompts him to leave and take up a fellowship at the University of Manchester**, England, in 1934. His mother stays behind and goes into hiding during Nazi rule

In 1938, Erdös accepts a year-long appointment at the **Institute for Advanced Study at Princeton**, where he **pioneers probabilistic number theory**

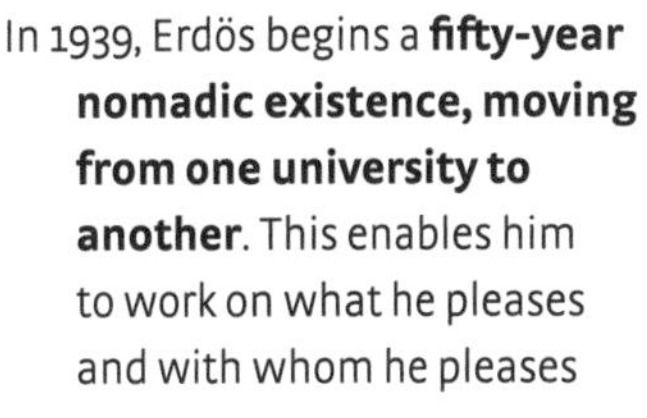

In 1939, Erdös begins a **fifty-year nomadic existence, moving from one university to another**. This enables him to work on what he pleases and with whom he pleases

He enjoys **collaborating with other mathematicians** and **inspiring young people**. Erdös would offer **cash prizes to those who could solve mathematical problems**

He is a force to be reckoned with, **working for twenty hours a day** and **producing around 1,500 mathematical papers**.

Erdös focuses extensively on **prime numbers**, **graph theory** (a field he helped to found), **set theory**, **geometry**, and **number theory**

In 1984, he wins the prestigious **Wolf Prize** and uses most of the $50,000 to establish a **scholarship in Israel**

Erdös **works right up to his death** on September 20, 1996. He spends the preceding hours grappling with a troublesome **geometric problem** at a **conference in Warsaw**

STEPHEN WOLFRAM

This brilliant mathematician and physicist was a child prodigy who became the brains behind the ground-breaking software package Mathematica and the Wolfram Alpha knowledge engine.

A CHILDHOOD GENIUS

Wolfram was **born in London on August 29, 1959**. Both his **parents were German Jewish refugees** who came to the United Kingdom.

Wolfram's genius shone in childhood. **By his early teens, he had compiled books on particle physics and a dictionary of physics**. He began publishing **scientific papers** from the **age of fifteen**.

He attended **St John's College, Oxford**, but **found the lectures "awful"** and left to join the **California Institute of Technology (Caltech)**, where he was awarded a **Ph.D. in particle physics at the age of just twenty**.

In 1973, Wolfram began using **computer experiments** to **lay the foundations of the emerging field of complex systems research**. This established **connections between computation and nature**, which Wolfram believed could be applied to **artificial life, weather prediction, stock market behavior**, and even the **origins of the Universe**.

His **early work** was mainly in **high-energy physics, quantum field theory**, and **cosmology**.

In 1981, he became the **youngest recipient of the MacArthur Fellowship at Caltech**.

Wolfram wrote up his ideas on **complex systems** in his 2002 book ***A New Kind of Science***. It became a best-seller.

MATHEMATICA

In 1988, Wolfram launched a **ground-breaking computer program** called **Mathematica**. It allowed the user to enter **complex mathematical equations**, which would then be **solved algebraically**. It was a hit in **education, engineering**, and **scientific research**.

$$$

Wolfram marketed Mathematica through his company **Wolfram Research Inc.**, and it made him extremely wealthy. The program has been updated many times since.

?

His latest innovations include **Wolfram Alpha**, a **search engine** that **delivers answers to specific questions** rather than a list of Web sites.

GRIGORI YAKOVLEVICH PERELMAN

This brilliant yet reclusive mathematician proved the Poincaré conjecture in 2003. He declined the Fields Medal and a $1 million dollar prize, preferring to live outside the limelight.

EARLY YEARS

Grigori Yakovlevich Perelman was **born on June 13, 1966, in Leningrad** (now St. Petersburg) in the former Soviet Union. His **mother studied mathematics** but **gave up postgraduate work to raise her son**.

Perelman excelled in mathematics from an early age and was enrolled at **Leningrad Secondary School no. 239**, which specialized in **mathematics** and **physics**. He won a **gold medal** at the **International Mathematical Olympiad** in 1982.

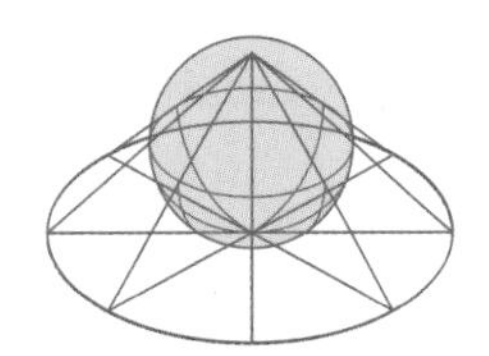

Perelman studied at **Leningrad State University School of Mathematics and Mechanics**, where he also gained his **Ph.D.** He went on to work at the **Steklov Institute of Mathematics** of the **USSR Academy of Sciences** before gaining **research posts at American universities**. He became known for work on **Riemannian geometry** and **topology**.

PROVING THE POINCARÉ CONJECTURE

In 2000, the **Clay Mathematics Institute** (CMI) offered **$1 million prizes for those who could solve key problems in mathematics**.

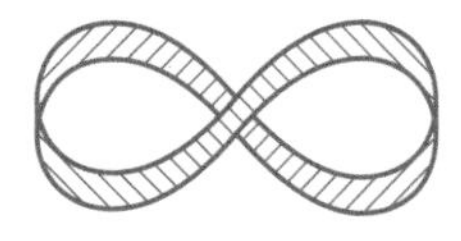

The Poincaré conjecture was among the CMI's **original seven Millennium Problems**. It was proposed in 1904 by the French mathematician **Henri Poincaré** and focused on **topology**.

The American mathematician **Richard Hamilton** developed the **geometric technique called Ricci flow**, beginning in 1981.

In 2003, Perelman **used Ricci flow to prove the Poincaré conjecture** and this was **independently confirmed**.

Three years later Perelman was awarded the **Fields Medal**, one of the highest accolades in mathematics, but he **declined it**.

The **CMI offered Perelman the $1 million prize in 2010**, but he **refused this, too**. It's **rumored that Perelman no longer studies mathematics**.

QUANTUM ALGORITHMS

Just as ordinary algorithms underpin the programs that enable everyday computers to solve problems, so quantum algorithms are designed to run on superfast quantum computers.

COMPUTING WITH QUBITS

An **ordinary binary digit**, a bit, can take the value 1 or 0. But a **quantum bit**, or **qubit**, is simultaneously a mixture of 1 and 0—thanks to the **weird laws of quantum physics**. And a **quantum byte**, made of **eight qubits**, can store 2^8, or 256 numbers simultaneously.

What this means is that a **quantum computer can perform 256 (and more) calculations in the time it takes an ordinary computer to do one**.

Rudimentary quantum computers have been built, and there's an **array of quantum algorithms to run on them**.

SHOR'S ALGORITHM

Devised by U.S. mathematician **Peter Shor**, this algorithm is **bad news for modern public-key encryption systems**. These rely on the **difficulty of factorizing very large numbers**. Shor's algorithm, however, can use a **quantum computer to factorize enormous numbers in polynomial time**. The good news is that **quantum systems bring with them their own impregnable encryption scheme**.

GROVER'S ALGORITHM

Like **Google Search** after fifteen espressos, Grover's algorithm—developed in 1996 by Indian-American computer scientist **Lov K. Grover**—makes **lightning-fast searches through unstructured data**. Whereas the **time taken by an ordinary search algorithm typically scales in proportion to the number of items to be searched through, *N*, Grover's algorithm scales as just $\sqrt{N}$**. So, if you increase the size of your data by a factor of 100, an ordinary algorithm will take 100 times longer to perform a search, while Grover's algorithm takes just ten times longer.

QUANTUM SIMULATION

Simulations of the **behavior of quantum mechanical systems**—the sort physicists might perform to develop their **theories of the subatomic world**—run far more efficiently on inherently quantum machines. The idea was first suggested in 1982 by physicist **Richard Feynman**, and is **now an active area of quantum computing research**.

ANDREW WILES

In 1995, 350 years after it was first stated, British mathematician Andrew Wiles proved Fermat's Last Theorem, succeeding where so many before had failed.

EARLY YEARS

On his way home from school in Cambridge in 1963, a **ten-year-old boy stopped at his local library**. He found a book about Fermat's Last Theorem and became **fascinated by this unproven conjecture**.

That boy was Andrew Wiles, who would **make history** by proving the theorem thirty years later.

WHAT IS FERMAT'S LAST THEOREM?

In 1637, Pierre de Fermat stated that **no three positive integers *a*, *b*, and c satisfy the equation $a^n + b^n = c^n$, where *n* is an integer greater than 2.**

PROOF OF FERMAT'S LAST THEOREM

Wiles has spent a lifetime studying mathematics. He earned a **B.Sc.** from **Merton College, Oxford**, in 1974, and a Ph.D. from Clare College, Cambridge. In 1982, he was appointed a **professor of mathematics at Princeton University**.

In adulthood, Wiles revisited his childhood dream of proving Fermat's Last Theorem. He **used many mathematical techniques**, including **algebraic geometry** and **number theory**, and **researched the theorem for seven years**.

He first presented his **129 pages of mathematical proof** in 1993; however, it **contained a mistake**. Wiles **rectified this** and **published his findings to international acclaim in 1995**. Finally, after more than 350 years, Fermat's Last Theorem had been proven.

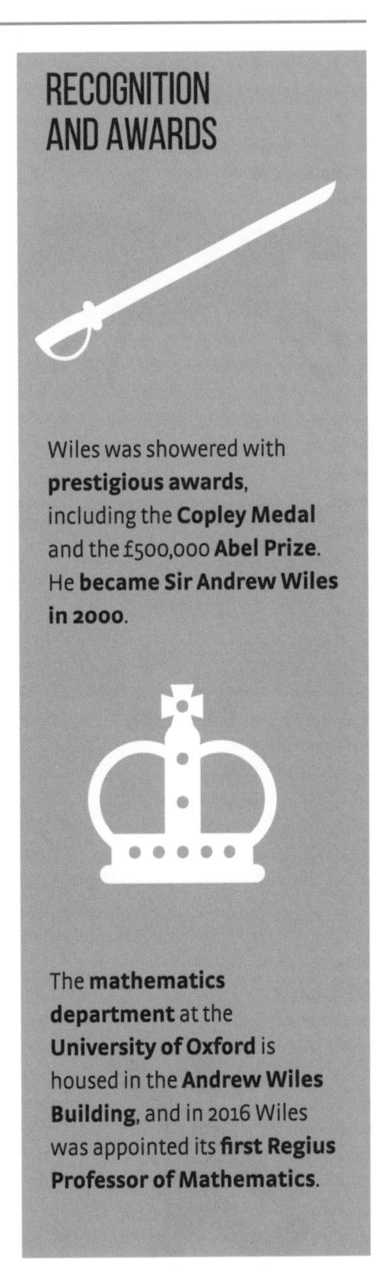

RECOGNITION AND AWARDS

Wiles was showered with **prestigious awards**, including the **Copley Medal** and the £500,000 **Abel Prize**. He **became Sir Andrew Wiles in 2000**.

The **mathematics department** at the **University of Oxford** is housed in the **Andrew Wiles Building**, and in 2016 Wiles was appointed its **first Regius Professor of Mathematics**.

SPORTS ANALYTICS

Just as early mathematicians supplemented their incomes playing cards and dice, so their modern counterparts have constructed mathematical models of sports in search of a gambling edge.

HORSE RACING

Perhaps the oldest sport for betting, **horse racing** is still one of the most popular among gamblers today. In the early 1980s, a physics dropout from Pittsburgh called **Bill Benter** set up a **sophisticated computer model of the sport**. It **pooled together statistics on horses' past performances** to spit out the **probability of any particular runner winning a race**. This enabled him to check **bookmakers' odds** and identify **value bets**. Benter started punting his model at the **race tracks of Hong Kong**, and to date is an estimated billion dollars richer for it.

BASEBALL

The analysis of **baseball statistics** has become known as **sabermetrics**, after the **Society for American Baseball Research (SABR)**. By the 1970s, aficionados were writing **computer simulations to implement their statistical models**. Perhaps most famously, **Billy Beane**, who was **general manager of the Oakland Athletics in the late 1990s**, used a form of sabermetrics to buy players who were **undervalued given their statistical record**. His story was told in the 2003 book ***Moneyball***, and the subsequent movie.

SOCCER

There's been a concerted effort to **analyze soccer** (British football) by treating goals as a **"Poisson process"—used previously to model** the arrival of random events like calls to a customer service center. In 1996, British statisticians **Mark Dixon** and **Stuart Coles** developed a model to **calculate ratings for each team from their past results**. Together with other factors, such as **home pitch advantage**, this enabled them to **calculate the probabilities of either team winning or the match ending in a draw**. The enormous volumes bet globally on soccer (billions of dollars per year) mean that **even a small edge can equate to handsome profits**.

MARYAM MIRZAKHANI

The first woman to win the Fields Medal in mathematics, Mirzakhani used advanced algebraic geometry to understand the structure and complexities of curved spaces.

EARLY LIFE

Maryam Mirzakhani was **born in Tehran, Iran, on May 12, 1977**. She grew up during the Iran–Iraq War.

In 1994, Mirzakhani became the **first female Iranian to win a gold medal at the International Mathematical Olympiad**.

She **studied mathematics at the Sharif University of Technology** and earned a **Ph.D. from Harvard University** in 2004. She was not a native English speaker and **took class notes in Persian**.

ACADEMIC WORK

After working at the **Clay Mathematics Institute** and **Princeton**, Mirzakhani became a **professor at Stanford University, California**, in 2009.

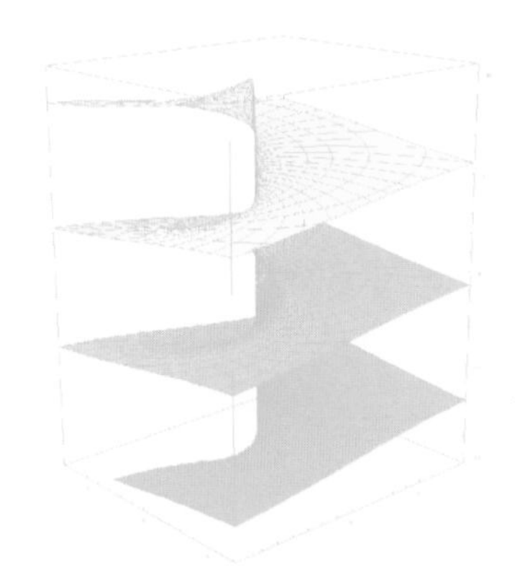

She was fascinated by the **moduli spaces of Riemann surfaces** and made contributions to understanding the **geometric and dynamic complexities of curved surfaces with simple loops** (i.e., loops that don't intersect themselves).

In her **doctoral thesis**, Mirzakhani **solved a highly complex problem regarding the number of simple loops (lines with the ends joined up) of a given length on a hyperbolic surface**. This led to **new proof of the Witten conjecture**, a theorem from **algebraic geometry** that has applications in **string theory**.

MATHEMATICAL HEROINE

In 2014, Mirzakhani became the first woman to receive the Fields Medal, the **highest mathematical honor in the world**. This was for her **"outstanding contributions to the dynamics and geometry of Riemann surfaces and their moduli spaces."**

Mirzakhani **would work on a problem** by **doodling** on large sheets of paper and **scribbling formulas** around her drawings. Her daughter described her mother's work as **"painting."**

Tragically, Mirzakhani was diagnosed as having **breast cancer** in 2013. The disease spread and she **passed away on July 14, 2017**, at the age of forty. Tributes flooded in from mathematicians across the world as well as from the **Iranian President Hassan Rouhani**.

TERENCE TAO

A child prodigy with an IQ off the scale, Terence Tao attended university from the age of nine. His ground-breaking theories place him among the world's greatest living mathematicians.

THE TAO OF MATH

Terence Tao was **born in Adelaide, Australia, on July 17, 1975**. By the **age of two he could teach other children how to read and count**.

His genius was evident at primary school and, **by the age of nine, Tao began studying mathematics part time at Flinders University, Adelaide**.

At the age of ten, he **became the youngest person to compete in the International Mathematical Olympiad**. Between 1986 and 1989, he won **bronze**, **silver**, and **gold medals**.

He **graduated at the age of sixteen**, earned his masters the following year and gained a **Ph.D. from Princeton at the age of twenty**.

Tao was **offered a lucrative post in a hedge fund but preferred to remain in academia**. He is a **mathematics professor at the University of California, Los Angeles (UCLA)**.

In 2004, Tao became famous after his **collaboration with English mathematician Ben Green**, proving that **there exist arbitrarily long arithmetic progressions of prime numbers**. This became known as the **Green–Tao theorem**.

In 2006, Tao received the prestigious **Fields Medal** for contributions to **partial differential equations**, **harmonic analysis**, and **additive number theory**. This prize is awarded to **exceptionally gifted mathematicians under the age of forty**.

Tao has been **showered with fellowships and prizes**, including the **Royal Medal** of the UK Royal Society, which he won in 2014 **"for his many deep and varied contributions to mathematics."**

A prolific author, Tao has **published more than 300 research papers and seventeen books**.

His **wife Laura is an engineer at the NASA Jet Propulsion Laboratory**. They live with their two children in California.

KAREN UHLENBECK

The first woman to win the prestigious Abel Prize, Uhlenbeck has pioneered new areas of mathematics, particularly in geometry, and she is an inspiration to female students across the world.

LIFE

Karen Uhlenbeck (**née Keskulla**) was **born on August 24, 1942, in Cleveland, Ohio**. She graduated from the **University of Michigan** and completed a **Ph.D. at Brandeis University, Massachusetts**, on **"The Calculus of Variations and Global Analysis."**

In the early 1980s, Uhlenbeck pioneered the mathematical discipline of **geometric analysis**, which is **the study of shapes using partial differential equations**.

Uhlenbeck has worked at a number of prestigious U.S. universities for many years and is now **Emeritus Professor of Mathematics at the University of Texas at Austin**. She is also a **visiting senior research scholar at Princeton**.

THE SCIENCE OF SOAP

One of Uhlenbeck's most exciting and important discoveries centered on **soap bubbles. Surface tension** pulls the bubbles into the shape with the **smallest possible surface area**, which ordinarily is a **sphere**.

Uhlenbeck studied the effects of **introducing other shapes into the soap** to see how the bubbles would react. She found that they formed a **network of planes meeting at certain angles. Adding higher-dimensional curved shapes created even greater complexity.**

Her findings have helped **theoretical physicists** develop tools to **understand the behavior of electromagnetic fields and quantum particles**.

BREAKING DOWN BARRIERS

The **Abel Prize**—one of mathematics' greatest accolades—was **awarded to Uhlenbeck** in 2019 for **"her pioneering achievements in geometric partial differential equations, gauge theory**, and **integrable systems**, and for the **fundamental impact of her work on analysis, geometry, and mathematical physics."**

INSPIRING WOMEN

As the first female winner of the Abel Prize, Uhlenbeck co-founded the **Women and Mathematics Program** at the **Institute for Advanced Study, Princeton**, with the **mission to recruit and retain more women in mathematics**.

DATA SCIENCE

Data science is a relatively new field that straddles mathematics, statistics, and computing. It is a practically motivated field that prioritizes prediction by the most effective means available.

PREDICTIVE MODELING

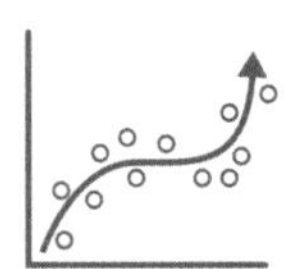

Traditionally the remit of **statistics**, using methods such as **regression analysis**, rapid **increases in computing power** have now **ushered in more sophisticated techniques**, such as **neural networks**, which might better be described as **machine learning** than statistics.

UNSUPERVISED MACHINE LEARNING

Most machine learning is **supervised**, meaning that the **algorithm is trained to predict a particular trait based on the input variables**.

But perhaps more interesting is **unsupervised learning**, where an **algorithm is let loose on a dataset with no prescribed brief** other than **"tell me something interesting about this data."**

Examples of unsupervised learning include **clustering algorithms**, which **sort data into groups**, and some **varieties of neural network**.

STATISTICAL COMPUTING

The term **big data** has also become **commonplace**. Indeed, **many statistical datasets used today are billions of rows in size**. And handling these means that those doing the analysis must be masters of some **heavy-duty computational resources**. These include:

Databases: these are used for **efficient storage and retrieval of data**. Common systems include **Oracle**, **Postgres**, and **MySQL**.

Data-analysis software: packages such as **R** and **Python** provide **ready-made implementations of common statistical and machine learning techniques**.

Processing power: skills in **cloud computing** and **parallel processing** help tap into the **raw computational power needed to crunch through massive volumes of data**.

DATA VISUALIZATION

As algorithms to analyze data become increasingly **obscure**, so the **sub-field of data visualization**—using **statistical plots** and **graphics**—becomes ever-more important. This **aids researchers in exploring data**, and also **helps to communicate findings in data science to the wider public**.

GOLDBACH CONJECTURE

It's one of the oldest, most well known, and perhaps the simplest to state, of all the unsolved problems in mathematics. Goldbach's conjecture is a theorem concerning prime numbers.

WHAT IS IT?

The **Goldbach conjecture was postulated in 1742 by the German mathematician Christian Goldbach**, in a letter to his colleague **Leonard Euler**. It says: **"Every even integer greater than 2 can be expressed as the sum of two primes."**

Goldbach numbers are commonly displayed in a **triangular diagram** with the even integers running down the center and their prime components along the outer edges.

So, for example, the first few even integers bigger than 2, expressed as sums of primes are 4 = 2 + 2, 6 = 3 + 3 and 8 = 5 + 3.

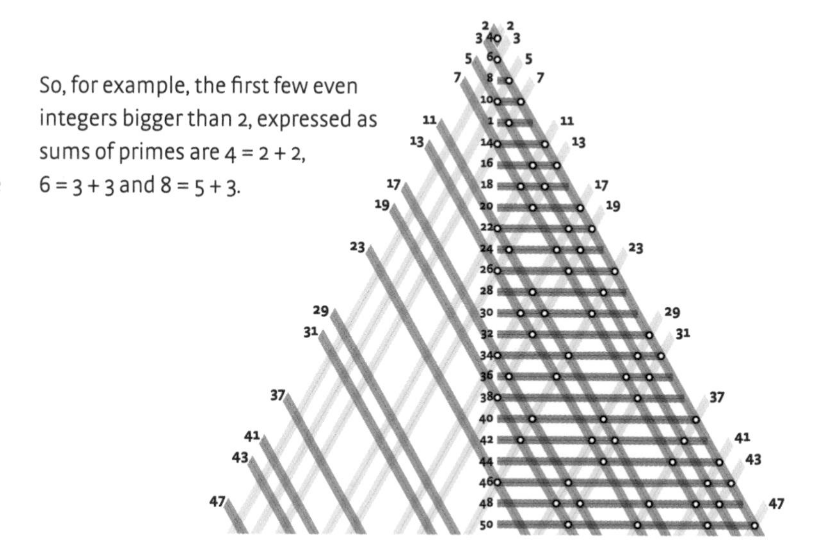

March 18, 1690 Christian Goldbach born in the Prussian city of Königsberg

1710 After studying at Königsberg's Royal Albertus University, he spends fourteen years traveling Europe

1725 Settles in St. Petersburg and becomes a professor of mathematics in the Academy of Sciences

June 1742 Proposes his conjecture, along with a number of variants, in a letter to the Swiss mathematician Leonhard Euler

November 20, 1764 Dies in Moscow

COMPUTER SOLUTION

Although **no formal mathematical proof of Goldbach's conjecture** has been produced, **Tomás Oliveira e Silva**, of the University of Aveiro, Portugal, has **organized a distributed computing project to check integers numerically**. The project has **verified the conjecture for all even integers up to 4×10^{18}**—that's 4 quintillion.

GLOSSARY

Analysis: the study of mathematical functions is known as analysis. It involves finding the minima, maxima, zeros, and how the function behaves at extreme values of the inputs.

Area: area gives the size of a 2-D surface. That might be a flat shape like a square or the outer boundary of a 3-D solid such as a cube. The area of a rectangle is just its width multiplied by height.

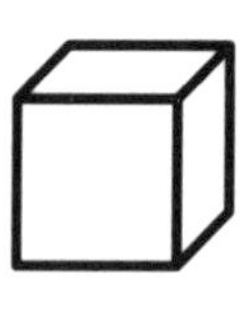

Arithmetic: the basic operations for combining numbers are known collectively as arithmetic. They are addition, subtraction, multiplication, and division.

Attractor: in mechanics, the long-term behavior that a dynamical system converges to for a wide range of starting conditions is known as an attractor. Chaotic systems have fractal attractors.

Bayesian statistics: when probability reflects our ignorance about something, such as tomorrow's weather, Bayesian statistics tells us how to update our beliefs based on the available evidence.

Binary numbers: binary, or base 2 arithmetic, is a counting system based around the digits 1 and 0. It is used in computing where 1 and 0 can be encoded as the on–off state of a switch.

Cipher: in cryptography, a cipher is an encryption system that involves replacing letters in a message—according to a secret set of rules—so as to render it unreadable by a third party.

Classical mechanics: mathematics applied to the laws of classical physics (i.e., those formulated prior to the introduction of quantum theory and relativity) is known as classical mechanics.

Complex number: a number that has both a real and an imaginary component, the latter of which is a multiple of $i(=\sqrt{-1})$, is called a complex number.

Deep neural network: a neural network is a machine-learning algorithm that mimics the layers of neurons in the human brain. A *deep* neural network consists of many such layers.

Denary numbers: base 10, or denary numbers, form a system of counting using the digits 0 to 9. Denary numbers are the counting system we are most familiar with.

Denominator: in a simple fraction, written x/y, the number y is known as the denominator. It gives the denomination of fraction, for example quarters ($x/4$) or thirds ($x/3$).

Differentiation: one of the two main branches of calculus, differentiation is concerned with finding the gradient, or rate of change, of a mathematical function with respect to its inputs. It's used extensively in physics.

Electromagnetism: the unified theory of electric and magnetic fields is known as electromagnetism and was formulated by physicist James Clerk Maxwell in 1873.

Ellipse: an ellipse is a closed curve given by the equation $\frac{x^2}{a^2}+\frac{y^2}{b^2}=1$, where a is the semi-major axis and b is the semi-minor axis. The special case $a = b$ corresponds to a circle.

Euler's identity: widely regarded as the most beautiful equation in all of mathematics, Euler's identity relates calculus, geometry and number theory in the expression $e^{i\pi} + 1 = 0$.

Exponential distribution: the exponential distribution is a continuous probability distribution for a random variable x, taking the form $p(x)=\lambda e^{-\lambda x}$, where λ is a free parameter of the distribution.

Factorial: the factorial of an integer, n, is used in combinatorics and probability theory. Denoted $n!$, it is calculated as $n \times (n-1) \times (n-2) \ldots 1$. For example, $3! = 6$ and $5! = 120$.

Fields Medal: awarded every four years to between two and four mathematicians under the age of forty, the Fields Medal is considered the mathematical equivalent of the Nobel Prize.

Frequentist statistics: frequentist statistics describe repeated trials of an inherently random process where the probability of each possible outcome is well established, like flipping a coin or rolling a die.

Geodesic: in the geometry of curved spaces, a geodesic is the shortest path between two points. In flat space, it's a straight line. On a sphere, it's a "great circle"—like the lines of longitude on Earth.

Gradient: the rate of change of a mathematical function with respect to its inputs is called the gradient. For example, if $y = 2x$ then the gradient is 2; for every increase in x,y increases twice as much.

Hyperbola: a hyperbola is the shape formed by cutting a vertical section through an upright cone. In the simplest case, it is given by the equation $y = 1/x$.

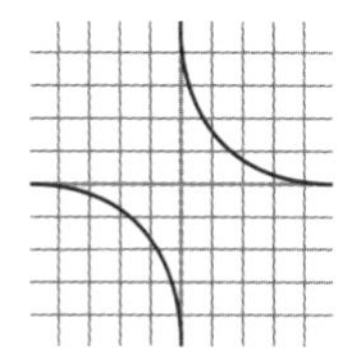

Hypotenuse: the longest side of a right-angled triangle is known as the hypotenuse. Pythagoras's theorem gives its length as the square root of the sum of the squares of the other two sides.

Information entropy: information entropy is a measure of the uncertainty associated with a random event. If the probability of the event is p, then the entropy is $-p \log p$.

Integer: a whole number with a zero imaginary component is known as an integer. The integers are governed by the branch of pure mathematics known as number theory.

Integration: integration is a sub-branch of calculus that amounts to the inverse of differentiation. Whereas differentiation calculates the gradient of a curve, integration gives the area enclosed beneath it.

Itô calculus: a method of applying the laws of calculus to analyze random walks, Itô calculus is used to study the behavior of stochastic processes such as movements of the stock market.

Machine learning: a branch of data science that employs computer algorithms to look for patterns in datasets, machine learning is used in everything from stock market prediction to music recommendation services.

Maximum likelihood: maximum likelihood is a method of fitting a statistical model to a dataset, by choosing parameter values that maximize the probability of the model producing the observed data.

Mean: also called the average, the mean of a random variable is the sum of all possible values that the variable can take, weighted by the probability of each value occurring.

Method of least squares: least squares is a technique for fitting statistical models to data, by tuning the model so that it minimizes the sum square differences between the model's predicted values and the data.

Millennium Problems: the Millennium Problems are a set of outstanding mathematical problems set out by the Clay Mathematics Institute in 2000. There's a $1 million prize for solving any of them.

Numerator: in a simple fraction, written x/y, the number x is known as the numerator. For example, *three*-quarters (3/4), or *two*-thirds (2/3).

Parabola: a parabola is a type of conic section. It is typically governed by equations of the form $y = ax^2$, where a is a constant. Projectiles fired into the air follow an approximately parabolic path.

Poisson distribution: the Poisson distribution is a discrete probability distribution for modeling the arrival of random events occurring at a constant rate. The probability of seeing k events is $p(k) = \frac{\lambda^k e^{-\lambda}}{k!}$, where λ is the mean number expected.

Posterior probability: the posterior probability, used in Bayesian statistics, is the optimal aggregation of our prior beliefs about an unknown quantity with new evidence.

Prior probability: in Bayesian statistics, the prior probability distribution encapsulates our knowledge about an unknown quantity *before* new evidence is taken into account.

Proportionality: if a quantity, y, scales in proportion to another quantity, x (for example $y = 2x$), then y is said to be *proportional* to x. Mathematicians denotes this as $y \propto x$.

Quadratic equations: equations involving a squared term in the unknown variable, x, are known as quadratic equations. They take the general form $ax^2 + bx + c = 0$. They can be solved either by factorization or by formula.

Quantum theory: the branch of physics dealing with tiny subatomic particles and the forces operating between them, quantum theory was born at the end of the nineteenth century.

Qubit: a quantum bit, or qubit, is the fuzzy unit of information used in the quantum theory of computation.

Relativity: the behavior of objects traveling close to the speed of light deviates significantly from the predictions of classical mechanics. Einstein's theory of relativity must be used instead.

Replication crisis: the replication crisis is the apparent failure to reproduce many results in the social sciences. It is thought to be due to flawed statistical analyses in the original studies.

Root: the inverse of raising a number to the nth power is called taking the nth root. In mathematical notation, the nth root of x can be written as $x^{1/n}$ or $\sqrt[n]{x}$.

Sexagesimal numbers: sexagesimal, or base 60, is a number system that was used by the ancient Babylonians. We still use it today in the measurement of time—for example, seconds in a minute or minutes in an hour.

Simplex method: the simplex method is an optimization algorithm that works by sorting through all possible solutions to a problem to find the best fit.

Simultaneous equations: a group of n equations in n unknown quantities are known as simultaneous equations. They can be rearranged to deduce the value of each unknown.

Sine: the simplest trigonometric function, the sine of an angle in a right-angled triangle is given by the length of the side opposite the angle divided by the length of the hypotenuse.

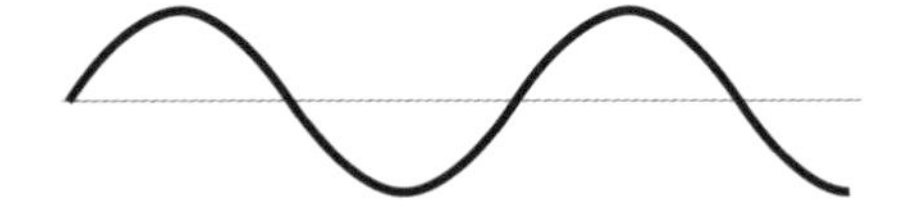

Standard deviation: the uncertainty in a random variable is given by its standard deviation, calculated by summing each value's squared distance from the mean, weighted by the probability of that value occurring.

Volume: the size of a 3-D object is measured by its volume. A cuboid, for example, has volume equal to its length multiplied by its height multiplied by its depth.

FURTHER READING

Alex Bellos, *Alex's Adventures in Numberland: Dispatches from the Wonderful World of Mathematics*

Tony Crilly, *50 Maths Ideas You Really Need to Know*

Keith Devlin, *The Millennium Problems: The Seven Greatest Unsolved Mathematical Puzzles of Our Time*

Richard Elwes, *Maths 1001: Absolutely Everything That Matters in Mathematics in 1001 Bite-Sized Explanations*

Richard Elwes, *Maths in 100 Key Breakthroughs*

Hannah Fry, *Hello World: How to be Human in the Age of Algorithms*

Martin Gardner, *The Colossal Book of Mathematics: Classic Puzzles, Paradoxes, and Problems*

Robert Kanigel, *The Man who Knew Infinity: A Life of the Genius Ramanujan*

Jim Al Khalili, *The House of Wisdom: How Arabic Science Saved Ancient Knowledge and Gave Us the Renaissance*

Robert Matthews, *Chancing It: The Laws of Chance and What They Mean for You*

Matt Parker, *Humble Pi: A Comedy of Maths Errors*

Simon Singh, *Fermat's Last Theorem: The Epic Quest to Solve the World's Greatest Mathematical Problem*

Simon Singh, *The Simpsons and Their Mathematical Secrets*

Ian Stewart, *Does God Play Dice? The New Mathematics of Chaos*

Ian Stewart, *Professor Stewart's Cabinet of Mathematical Curiosities*

Y
Z
X